CAROLYN FRY

SEEDS

Safeguarding Our Future

IVY PRESS

For my mother and father,
Jane and Roger Fry

First published in the UK in 2016 by
Ivy Press
210 High Street
Lewes
East Sussex BN7 2NS
United Kingdom
www.ivypress.co.uk

British Library Cataloguing-in-Publication Data
A catalogue record for this book is available from the British Library

ISBN: 978-1-78240-324-1

This book was conceived, designed and produced by
Ivy Press

PUBLISHER Susan Kelly
CREATIVE DIRECTOR Michael Whitehead
EDITORIAL DIRECTOR Tom Kitch
ART DIRECTOR Wayne Blades
COMMISSIONING EDITOR Jacqui Sayers
PROJECT EDITOR Joanna Bentley
BOOK DESIGN Evelin Kasikov
ILLUSTRATOR Emma Kelly
PICTURE RESEARCHER Katie Greenwood

Printed and bound in China

10 9 8 7 6 5 4 3 2 1

Front cover: Light micrograph of strawberry seeds.
Science Photo Library/Eye of Science.

Back cover (top to bottom): Getty Images/Darlyne A. Murawski/National Geographic; Royal Botanic Gardens, Kew; Shutterstock/Ethan Daniels; Shutterstock/Saied Shahin Kiya; Science Photo Library/Power and Syred.

CONTENTS

8 Seed Banks Around the World

10 Introduction

CHAPTER 1

14 **The Importance of Seeds to Humanity**

16 *Separating Humans from the Monkeys*

18 *From Hunter-Gatherers to Farmers*

26 *How Crop Wild Relatives Have Helped Us Breed Resilient Varieties*

28 *Human Uses of Seeds Down the Ages*

30 *The Father of Seed Science*

32 *The Seed Bank that Survived a Siege*

36 *Plants and Seeds from the World's Arid Lands*

40 *Seed Profile: Grass Pea*

CHAPTER 2

42 **How Plants Evolved on Planet Earth**

44 *Tiny Algae Give Rise to the First Plants*

48 *Spore-Bearing Plants Give Rise to the First Seeds*

52 *Flowering Plants Quickly Gain Ground*

58 *The Rise of Annuals Underpins Human Success*

60 *How Plants Evolved from Algae to Angiosperms*

62 *Evolution of Land Plants Culminates with the Dramatic Rise of Angiosperms*

64 *Plants and Seeds from the World's Rainforests*

68 *Australia's Plantbank Helps to Grow "Difficult" Rainforest Seeds*

72 *Seed Profile: Wollemi Pine*

CHAPTER 3

74 How Seed Plants Reproduce

76 *Double Fertilization Brings Flowering Plants Great Success*

82 *How Gymnosperms and Angiosperms Reproduce*

84 *Pollination is a Must for Successful Fertilization*

92 *Plants and their Pollinators*

94 *Saving China's Diverse Flora*

98 *Plants and Seeds from Antarctica and the Arctic*

102 *Seed Profile: Yew*

CHAPTER 4

104 Dispersal Takes Seeds to New Pastures

106 *The Diverse Ways in which Plants Spread their Seeds*

110 *Animals Disperse Seeds Far and Wide*

114 *Floating Down Rivers and Across the High Seas*

116 *Hitching a Ride with the Wind*

119 *Dispersal by Gravity and Ballistic Propulsion*

120 *How Seeds are Dispersed Around the World*

122 *The Seed Bank Keeping New York City Green*

128 *Plants and Seeds from the World's Islands*

132 *Seed Profile: Mongongo*

CHAPTER 5

134 Germination Brings Plants Back to Life

136 *The Tricks Plants Use to Survive*

144 *The Test of Time*

148 *Preserved for Posterity*

150 *Inside a Seed*

152 *Inspiring Seed Banking Around the World*

156 *Plants and Seeds from the World's Coastal Zones*

158 *Seed Profile: Wood Anemone*

CHAPTER 6

160 Using Seeds to Ensure Humanity's Survival

162 *Saving Cultivated and Wild Seeds*

166 *Healthy Ecosystems are Key to Biodiversity*

168 *Restoring Biodiversity to Chalk Downlands*

172 *The Shifting State of the World's Flora*

174 *Keeping Hunger at Bay in the Tropics*

178 *Foods of the Future*

180 *Plants and Seeds from the World's Alpine Habitats*

184 *Seed Profile: Arabica Coffee*

186 Glossary

187 Further Reading

189 Picture Credits

190 Index

192 Acknowledgments

SEED BANKS AROUND THE WORLD

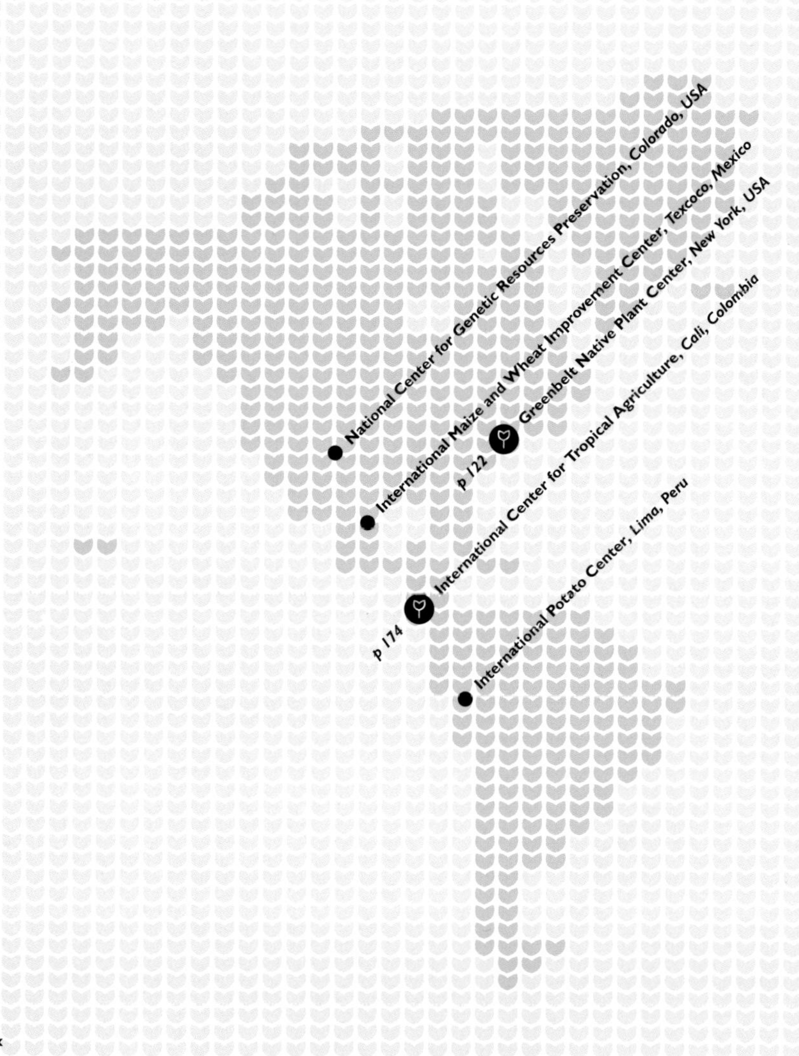

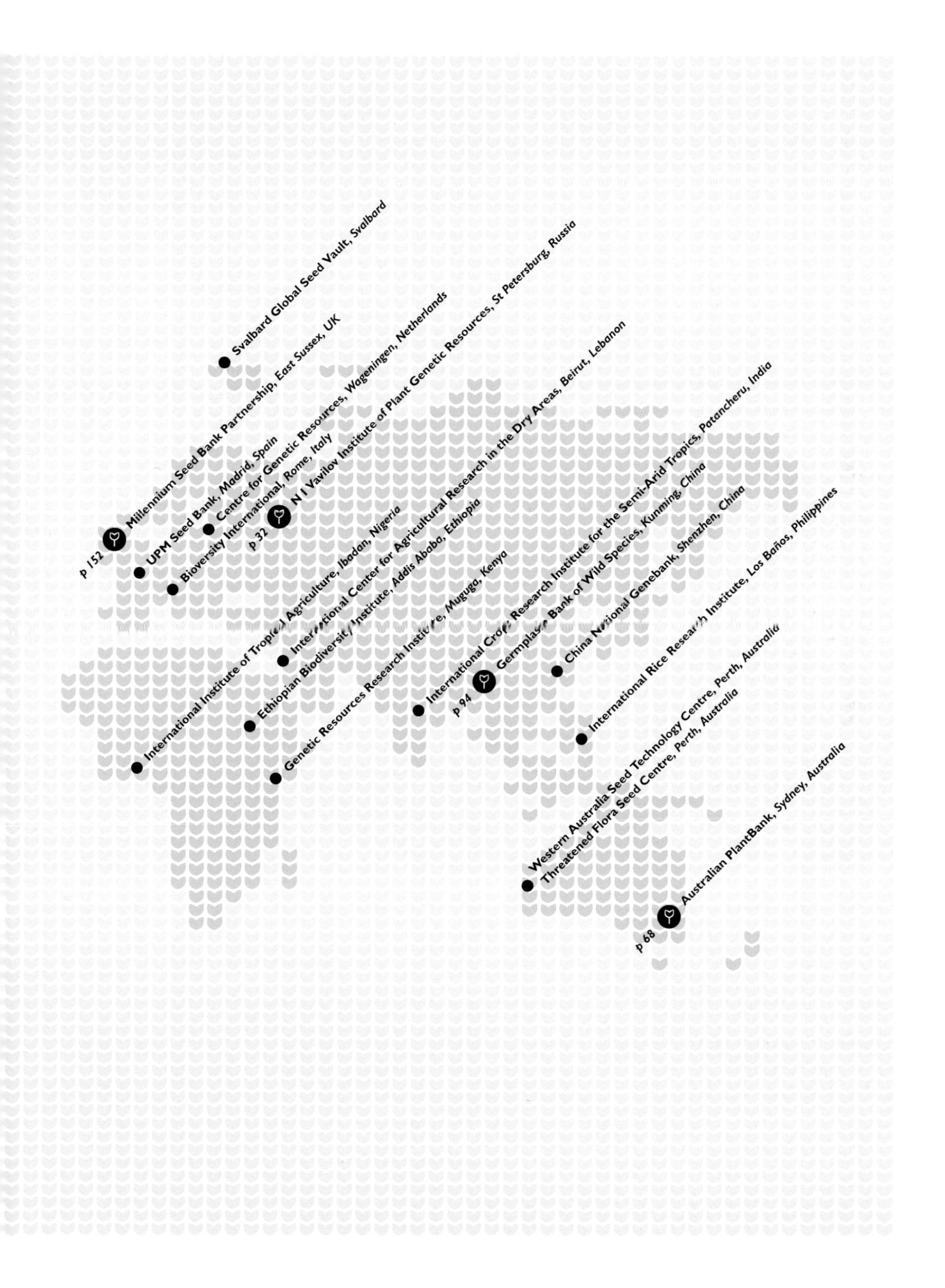
Svalbard Global Seed Vault, Svalbard
Millennium Seed Bank Partnership, East Sussex, UK
p 152
UPM Seed Bank, Madrid, Spain
Centre for Genetic Resources, Wageningen, Netherlands
Bioversity International, Rome, Italy
N I Vavilov Institute of Plant Genetic Resources, St Petersburg, Russia
p 32
International Center for Agricultural Research in the Dry Areas, Beirut, Lebanon
International Institute of Tropical Agriculture, Ibadan, Nigeria
Ethiopian Biodiversity Institute, Addis Ababa, Ethiopia
Genetic Resources Research Institute, Muguga, Kenya
International Crop Research Institute for the Semi-Arid Tropics, Patancheru, India
Germplasm Bank of Wild Species, Kunming, China
p 94
China National Genebank, Shenzhen, China
International Rice Research Institute, Los Baños, Philippines
Western Australia Seed Technology Centre, Perth, Australia
Threatened Flora Seed Centre, Perth, Australia
Australian PlantBank, Sydney, Australia
p 68

INTRODUCTION

As this book sets out to demonstrate, we have much to thank seeds for. Not only do they provide half the calories consumed by humanity today, but they also helped humans to evolve and civilization to develop. Some 3.5 million years ago, seeds provided a calorific boost to early hominids, who moved out of the trees and on to Africa's savannas; switching from eating leaves and fruit to consuming the seeds of grasses and sedges helped them develop bigger brains and gave them an evolutionary advantage over other primates. It was these early hominids, our ancient ancestors, who gave rise to the human race.

Later, the seeds of annuals, such as maize, wheat, and rice, helped to underpin the rise of civilization. Because they have a yearly lifecycle, annuals can be cultivated relatively easily. Once early humans discovered they could take control of their food supply by farming plants, they were able to swap a nomadic lifestyle for a more settled existence in ever-larger communities. In the majority of cases, human societies grew up on seed-based diets, although roots and tubers were also important. Not having to spend their entire time foraging enabled these early citizens to do other things. They began bearing more children, organizing their communities in a structured way, and studying the world around them.

The story of seeds begins long before humans came on the scene, however. To understand where seeds and plants come from requires us to go back 450 million years, some 220 million years before the dinosaurs evolved. At this time, ocean-dwelling algae that had evolved over time from single-celled microbes began inhabiting pools of freshwater on land. Slowly, a process of natural selection began, whereby individuals that could tolerate the surrounding environmental conditions survived, and those that could not fell by the wayside. Over many millions of years, as plants competed with each other for space, sunlight, water, and nutrients, a complex series of evolutionary changes led to the great diversity of land plants we see on Earth today.

The evolution of the seed proved to be particularly beneficial for plants. Before seeds, plants reproduced by spores, a process requiring water. Initially, plants produced spores of one size in their 'sporangia'

A new plant emerges from a seed at New York's Greenbelt Native Plant Center, one of the many seed banks that are helping to preserve biodiversity.

but later they evolved to produce large spores that became female reproductive organs and small spores that became male reproductive organs. Eventually, the female parts of sporangia went from producing many spores to creating just one, and retaining this spore on the plant instead of dispersing it. This became the seed, while male spores became pollen. Not requiring water to reproduce meant seed-bearing plants could tolerate drier conditions, a great advantage as climatic conditions changed.

Today, two types of seed-bearing plants exist. Gymnosperms, which include conifers, number around 1,000 species, while angiosperms, the flowering plants, number 350,000 species. The angiosperms diversified very rapidly after evolving around 100 million years ago, a process described by naturalist Charles Darwin in 1879 as an "abominable mystery." He had considered that evolution only took place via natural selection in small, gradual steps. Scientists are still uncertain exactly what caused such a rapid diversification, but it is likely to involve flowering plants' innovative method of sexual reproduction, their relationships with animal pollinators, and the existence of a wide variety of climatic conditions throughout their evolution.

What is certain is that the evolution of angiosperms provided plentiful plants that enabled humans to thrive. All the major root crops and vegetables we eat are angiosperms, although we rely on less than ten crops to provide 75 percent of human energy needs. These are: rice, maize, wheat, sorghum, millet, potatoes, sweet potatoes, soybean and sugar from cane and beet. Seeds are our primary source of calories and protein, with those from cereals being the most important.

Seeds from wild plants are much like people; they each have a unique genetic profile. Over the thousands of years that humans have cultivated plants for food, they have selectively bred out traits that were not helpful for farming and bred in traits that were favorable. In generally stable climates, such as those experienced by farmers until recently, the ability to survive weather conditions outside the normal range was not that important. While traits such as retaining seeds on the plant for longer, resisting disease, and providing an attractive color and pleasant flavor were desirable. This essentially meant that farmers retained genes that gave favorable results and discarded those that did not. In doing so, they developed landraces that were highly suited to local conditions.

As farming grew in scale, and techniques for creating new cultivars became more complex, the genetic diversity within many crop plants was whittled down to the bare minimum required to give high yielding, good tasting, evenly ripening crops. Growing crops with the same or similar genetic profile meant farmers knew exactly what they were going to reap, but it left the crops vulnerable to disease and climatic changes. In the mid-19th century, Irish farmers learned the hard way about the downside of farming crops with no genetic diversity, when a water mould killed the nation's entire potato crop. And farmers today are learning of the importance of genetic diversity in growing crops that are resilient to climate change.

The environmental damage caused by humans to the natural world has become increasingly apparent over the past century. The Millennium Ecosystem Assessment uncovered the vast extent of the problem, concluding that: "Over the past 50 years, humans have changed ecosystems more rapidly and extensively than in any comparable time in human history, largely to meet rapidly growing demands for food, fresh water, fiber and fuel. This has resulted in a substantial and largely irreversible loss in the diversity of life on Earth." Faced with growing populations, increasingly infertile land, scarce water resources, and unpredictable climates, humanity is now facing a fight for survival.

In the past three decades, as concerns about the state of the planet have intensified, many seed banks have been established around the world. Essentially gene stores, they conserve the genetic diversity

*Seeds of Queen Anne's lace (*Daucus carota*). Genes from this wild carrot may help to make the cultivated carrot more resilient to climate change.*

present in wild plants, as well as that within agricultural landraces. They offer the chance to: breed diversity back into genetically sparse cultivars and make new crops that are resilient to climate change; support in situ conservation by providing seeds and plants to supplement depleted biodiversity; and prevent rare plants that may be valuable to humanity from going extinct. Each chapter of this book features a profile of one of the world's major seed banks, giving an insight into the essential conservation work they do.

This book sets out to tell the story of seeds and of how they have helped humans thrive on Earth. It takes a journey through their evolutionary history, reveals the complex process of plant reproduction that gives rise to them, uncovers the tricks they employ to ensure they germinate at the optimal time, and demonstrates the vital part they have played in creating biodiverse ecosystems. It explores the stories behind some of our most fascinating and useful seeds, and highlights the ways in which plants and their seeds adapt to different habitats. Looking to the future, it shows how modern scientific techniques of genetic profiling, seed banking, and plant breeding might contribute to overcoming some of the damage humanity has inflicted on the planet. Having helped us to become the most successful species on Earth, seeds may also offer our best hope for saving us from ourselves.

CHAPTER 1

THE IMPORTANCE OF SEEDS TO HUMANITY

SEPARATING HUMANS FROM THE MONKEYS

Without seeds, Western civilization might never have arisen, for these energy-giving capsules of life have underpinned our evolutionary history over millions of years. Specifically, it was annuals, those flowering plants that produce seeds once in the course of a year and then die, which fed the earliest settled communities of Mesopotamia, the cradle of Western civilization.

Today, the three most important crops in the world, providing about 60 percent of the world's food energy intake, are all annuals from the grass family: maize, wheat, and rice. These plants' seeds are packed full of the essential carbohydrates, protein, vitamins, and minerals that humans need to survive, while their fast life cycles make them easy for us to cultivate in large numbers. Although tubers, such as manioc and potatoes, have also played an important role in developing human societies, without seeds to eat, Earth might not have been able to support such a large human population.

WALKING TO A BETTER DIET

The animals that gave rise to humans did not initially eat seeds, however, our far-distant ancestors lived an apelike existence in forests, foraging leaves and fruits from trees, shrubs, and herbs. But 3.5 million years ago (Ma), these early hominids evolved from being tree climbers to upright walkers and moved onto the savannas. As they did so, their diet changed to reflect these new surroundings, becoming richer in grasses and sedges, and supplemented occasionally with meat, as demonstrated by evidence from the analysis of teeth taken from the remains of early species. Being able to eat a greater variety of foods in turn enabled these hominids to live in a wider range of ecosystems, so they were no longer competing for food with other primates. This advantage helped set humans on the path to becoming the most successful species of the animal kingdom.

FOOD FACTORS IN EVOLUTION

Studies into the nutritional quality of foods eaten by apes and hunter-gatherers have concluded that the leaves of wild plants are low in energy but high in minerals and vitamins, while seeds, nuts, roots, and tubers have a net

1 *Evolution of the human skull over time; from (far left)* Australopithecus africanus *(3–1.8 Ma) to (far right) the early modern human, Cro-magnon, (22,000 years ago).*

2 *The diet of early hominids shifted from comprising mostly leaves and fruit to including grasses, sedges, and meat.*

EATING NUTS ROTTED HUNTER-GATHERERS' TEETH

Acorns, the single-seeded fruits of oak trees, were a staple food of hunter-gatherers as far back as 21,000 years ago. However, they may also have rotted their teeth. A study of charred plant remains from an archaeological site in Morocco showed that the inhabitants were harvesting and processing sweet acorns and pine nuts between 13,700 and 15,000 years ago. The teeth of 52 skeletons buried in a cave in Taforalt exhibited a high level of tooth decay; only three of the skeletons had no cavities. Indeed, grinding the nuts into a sticky porridge before eating them might have helped to promote the disease. Prior to this study, scientists had thought that tooth decay only arose with the emergence of agriculture, when eating sugary wheat and barley became common.

energy yield that is higher than leaves but lower than meat. Today, apes and monkeys primarily eat leaves, which are of lower quality and easy to acquire, while hunter-gatherers mostly eat higher-quality seeds, nuts, tubers, and meat, which are harder to come by.

The shift from the ape diet to the hunter-gatherer diet 3.5 million years ago represents a move from lower-quality foods to higher-quality ones. Brains and guts both require considerable energy to function, and while apes have small brains and large guts, modern humans have large brains and small guts, an evolutionary transition that scientists believe would have been facilitated by the shift to eating higher-quality food.

COOKING CONTRIBUTES TO HUMAN DEVELOPMENT

How hunter-gatherers prepared their food, such as by grinding seeds and cooking them, might also have contributed to their evolution, as breaking down then cooking seeds and nuts would have made them quicker to eat and easier to digest. This would have enabled the early hominids to consume more calories and spend less time chewing, which could have been a contributory factor in the evolution of larger brains, bigger bodies, and smaller teeth. The brain of our ancestor *Homo erectus*, who first appeared 1.6 to 1.9 million years ago, was 50 percent larger than that of its predecessor, *Homo humilis*, and its teeth were also much smaller. A bigger brain would have enabled members of the species to process more complex information as they traversed locations rich in food sources. Meanwhile, a bigger body would have helped them travel farther on a day-to-day basis; *Homo erectus* was the first hominid to have a geographic range extending beyond a single continental region.

The first convincing evidence for cooking is dated to 790,000 years ago, after the era of *Homo erectus* but still some 600,000 years before the emergence of modern humans, *Homo sapiens*. Archaeologists examining the remains of a settlement at Gesher Benot Ya'aqov, in Israel, found a small number of charred remnants as they sifted through thousands of seeds and fragments of wood and fruit, from which they concluded that the proportion of burnt remains would have been greater had the materials been exposed to uncontrolled wildfires. They also found clusters of burnt flint, which they interpreted as primitive hearths. Meanwhile a study of modern-day hunter-gatherers has shown that no current human group exists that eats all its food raw.

FROM HUNTER-GATHERERS TO FARMERS

The Fertile Crescent around 8,000 years ago. At this time, the transition from foraging to farming was in its early stages. Over time, agricultural and pastoral economies spread northwest across Asia Minor into Europe, southwest to the Nile Valley and east towards central Asia and the Indus valley.

Hunter-gatherer communities switched independently of one another from foraging for wild seeds to cultivating them in several locations from 3,000 to 12,000 years ago. The earliest transition occurred in the Fertile Crescent area of Western Asia, a wide arc spanning parts of modern-day Israel, Lebanon, Jordan, Turkey, Iraq, and Iran. Early agriculture in this area was based around cultivating barley and wheat as these were the crops that were locally available.

Elsewhere the shift from hunter-gathering to farming was based on whichever plants were growing nearby. In China, this was rice and millet; in Papua New Guinea, root and tree crops were planted; sub-Saharan Africans grew sorghum and pearl millet; in Mesoamerica, maize and beans were the main crops; several seed plants were grown in eastern North America; and South Americans cultivated quinoa and beans. Recent evidence suggests that this process was a long, complex evolution, which involved some people cultivating crops long before settling in fixed locations and others farming settled communities while still actively foraging.

WHY DID EARLY HUMANS SWAP FORAGING FOR FARMING?

The shift from foraging to farming spanned the end of the last ice age, some 10,000 years ago. Some scientists believe the change in climates and atmospheric conditions may have contributed to this transition. Evidence from ice cores shows that ice-age climates were dry but also variable over short timescales, and that carbon dioxide (CO_2) in the atmosphere was low. Even though populations were relatively sophisticated at this time, the conditions were not conducive to agriculture.

At the end of the ice age, climates became warmer, rainfall increased, and concentrations of CO_2 rose. Temperate grass species, which include the ancestors of many major domesticated crops, would have thrived under these conditions. So, the scientists reason, once the conditions for agriculture became viable, different hunter-gatherer communities quickly exploited the opportunity this presented for taking control of their food supplies. Most present-day hunter-gatherers exist in areas that are either too arid or too cold for farming to be viable.

POPULATION BOOM

Shifting from a society that ate solely what nature provided to one that had some control over its food supply had considerable implications for early humans. Prime among these was that such a change promoted population growth. Women in foraging societies spend up to four years breast-feeding their children, an act that suppresses ovulation and so prevents more frequent births. This makes sense in the context of a mobile lifestyle, as carrying and feeding more than one child while hunter-gathering would be difficult. Once farming provided a more assured food supply, this enabled large communities to settle in one place, and as a result fertility increased and populations began to rise. At the end of the last ice age, the global population is estimated to have been five million. By 1820, it had swelled to one billion (1,000 million).

Historically, as more food was needed to feed a region's growing population, new land was simply brought under cultivation. By the nineteenth century, however, the most fertile land was becoming scarce, so the only option was to cultivate poorer-quality land. Many academics at this time were pessimistic about the world's ability to feed its population in the years to come, particularly in the light of longer life expectancies. The economist Thomas Malthus forecast that: "The power of population is so superior to the power in the Earth to produce subsistence for man, that premature death must in some shape or other visit the human race." Malthus had not bargained on advances in agricultural and seed science, however. In the twentieth century, large public investments in scientific research for agriculture brought about huge yield increases in developed countries.

Gramineae.
14. Olyreae.
1
2
3
A
4
B
82. Zea Mays L.
Gemeiner Mais.
I

1 *Maize, a member of the grass family, helped to support the shift from hunter-gathering to farming in Mesoamerica.*

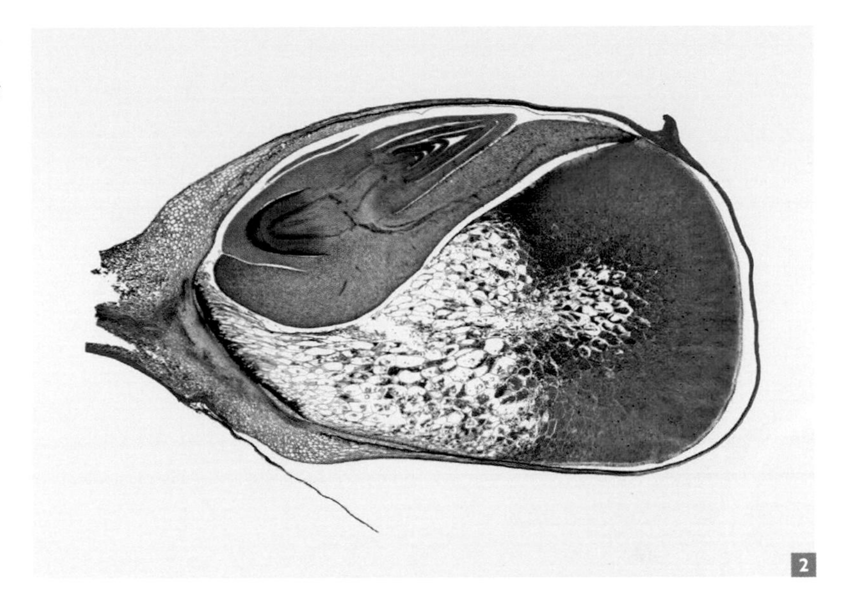

2 *This micrograph shows a maize kernel. The embryonic plant is the green section (upper left), while its food supply, called the endosperm, is the pink and multi-colored section (from center left to right). Two thirds of all human calories come from endosperm within seeds.*

NEW TECHNIQUES BOOST YIELDS

The rise in English wheat yields exemplifies the advances made at this time. Although it had taken nearly 1,000 years for wheat yields in England to rise from ⅕ to ⅘ of a ton per acre (0.5 to 2 metric tons per hectare), it took just 40 years to increase them from ⅘ to 2½ tons per acre (2 to 6 metric tons per hectare). The improvements came about thanks to innovative plant-breeding techniques, the development of inorganic fertilizers, and modern pesticides. These advances enabled most industrialized nations to attain food surpluses by the second half of the twentieth century. When India suffered recurring droughts in the mid-1960s, which threatened wide-scale starvation, the same methods were employed to increase rice and wheat yields there. This involved creating varieties that were more responsive to soil nutrients; had shorter, stiffer stems that could support the weight of heavier ears of grain; and could grow at any time of the year, enabling farmers to sow more crops on an annual basis.

Within five years, around 20 percent of wheat areas and 30 percent of rice areas were growing high-yield varieties; by 1990 this was 70 percent of each. Farmers who could make more profit from these varieties expanded their cultivation at the expense of other crops. Cereal production in Asia doubled from 1970 to 1995, while the population rose by 60 percent.

ANCIENT SEEDS AND FRUITS REVEAL SHIFT TO FARMING

Some of the best evidence of the transition from hunter-gathering to agriculture comes from the archaeological site of Abu Hureyra in Syria. This village of several hundred inhabitants was permanently inhabited from 7,000 to 11,500 years ago. The site yielded 118 species of seeds and hard fruits that would have been eaten by hunter-gatherers during the site's early days. However, the array of wild plants declined rapidly around 11,050 years ago, to be replaced 9,860 years ago by a suite of cultivated crops, including einkorn wheat, emmer wheat, and lentils. By 8,500 years ago, Abu Hureyra's inhabitants were relying on a mere eight domesticated plants for their vegetable-based energy foods.

Far from starving, the people enjoyed 30 percent more calories per person than they had previously. A downside to the benefits brought by this "Green Revolution," however, was the environmental damage it caused: fertilizers polluted waterways; poor irrigation practices caused salt to build up in the soil, rendering previously good-quality land unusable; and the heavy dependence on a few varieties reduced biodiversity. Perhaps most importantly, breeding within a limited set of popular varieties that favored high yields reduced their overall genetic diversity.

LOSS OF GENETIC DIVERSITY

A handful of wild wheat seeds is genetically diverse in much the same way as a crowd of people. Just like humans, plants in different wild populations can have markedly different characteristics despite being the same species. When farmers began to cultivate crops, the first seeds they sowed from wild plants would have contained only a small subsection of the genetic diversity present in their local wild wheat population. This, in effect, created a bottleneck in genes at the point at which agriculture first developed. Over subsequent millennia, farmers domesticated plants through a process of selection and breeding. They bred out natural traits, such as the shattering of seed heads and dormancy, which enabled plants to survive in the wild but were not useful for agriculture. On the other hand, they retained and selected for characteristics that were helpful, such as higher yields and pleasant taste. Any individual landrace is therefore the result of breeding work by thousands of farmers over many generations.

Modern cultivars are the result of sophisticated programs of breeding and genetic improvement specifically designed to meet the needs of large-scale commercial agriculture. Monoculture farming, under which vast areas are planted with a few such cultivars, results in the highest yields but the lowest genetic diversity. The Green Revolution is an example of this. At the start of the twentieth century, India was home to over 30,000 varieties of rice; today, just ten varieties are grown in 75 percent of the country's rice fields. Moreover, some crop varieties have been genetically modified to tolerate specific herbicides such as Monsanto's Roundup product, a development that has enabled farmers to easily eradicate weeds but has led to them abandoning traditional landraces and varieties in favor of crops that offer better financial returns.

INTERNATIONAL TREATY HELPS KEEP FOOD ON THE TABLE

Signed by 135 countries, the International Treaty on Plant Genetic Resources for Food and Agriculture came into force in 2004. Known as the International Treaty, it aims to ensure that farmers and breeders have access to the plant genetic resources they need, including seeds of food and forage crops, to overcome future challenges to farming such as climate change and environmental issues. The Treaty recognizes that all countries depend for food and agriculture on plant genetic resources that originated elsewhere. For example, wheat originated in western Asia but is now grown most extensively in China and India; and potatoes came from South America but are widely grown in Europe. The Treaty's key aim is: "The conservation and sustainable use of plant genetic resources for food and agriculture and the fair and equitable sharing of the benefits arising out of their use, in harmony with the Convention on Biological Diversity, for sustainable agriculture and food security."

PLANTS UNDER PRESSURE

How a crop plant responds to stresses, such as extremes of temperature, drought, and attacks from pathogens or pests, is at least in part controlled by that plant's genes. When a field of a particular crop is genetically diverse, that population has greater resilience, as individual plants respond to the stress in different ways. But when an entire population has the same genetic makeup, it does not have the armory to resist a particular stress or attack to which it is vulnerable. With climate change increasing the frequency of droughts and floods and making weather in many locations more variable, the concern is that modern-day cultivars may not be sufficiently resilient to grow the volumes of food required in future. This is of huge concern, as 80 percent of humanity's calorie intake now comes from just 12 plant species, while half of our calories come from the three grasses: maize, wheat, and rice. If these are vulnerable, the impacts of their loss will be huge.

1 *Wheat grains, seen here, are ground down to make flour for food products, such as bread and pasta.*

2 *Wheat, one of the world's most important crops, ripens in a field on rolling chalk downland in England.*

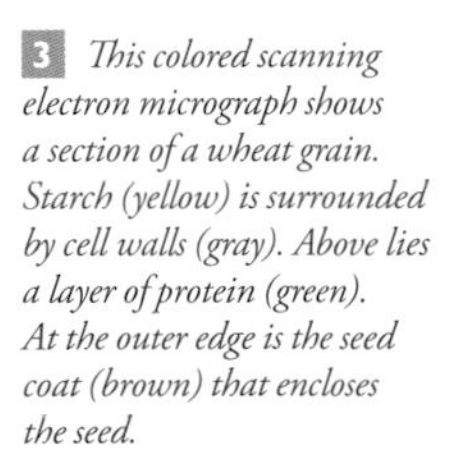

3 *This colored scanning electron micrograph shows a section of a wheat grain. Starch (yellow) is surrounded by cell walls (gray). Above lies a layer of protein (green). At the outer edge is the seed coat (brown) that encloses the seed.*

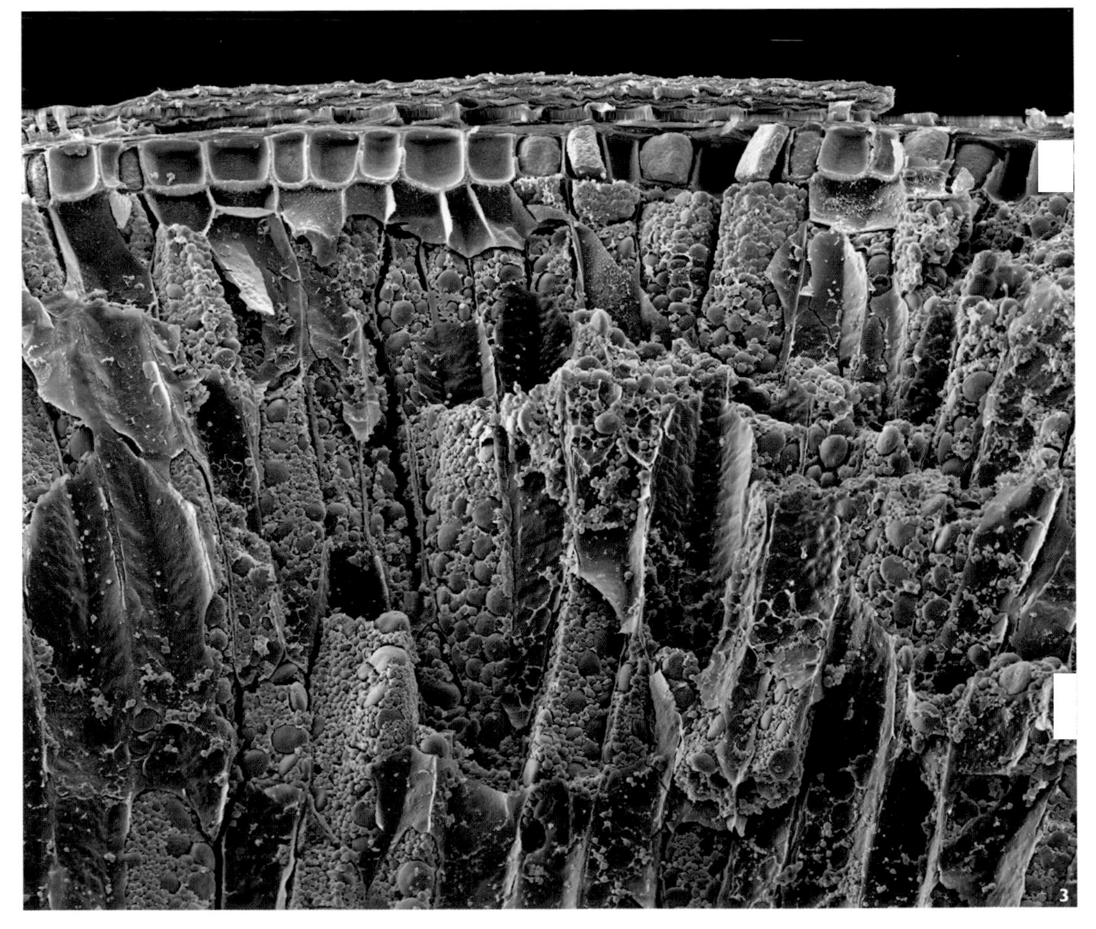

1 **2** *Scientists are hoping to enhance the resilience of crops such as lentils (above) and sorghum (below right) through research into crop wild relatives.*

DIVERSITY ENSURES RESILIENCE

There is already evidence that resistance is weakening. Today, the world's fourth most important crop after wheat, rice, and maize is the banana. However, nearly all commercially grown banana plants are a cultivar called Cavendish; they are all clones, which makes them genetically identical. Since 1992, 24,700 acres (10,000 hectares) of Cavendish plantations have succumbed to a strain of Panama disease fungus. With no genetic diversity to fight the disease, the entire global crop is at risk. "Plants with little genetic diversity are very vulnerable because there's no variability there," explains Ruth Eastwood of Kew Gardens' Millennium Seed Bank Partnership (MSBP). "If one field of one crop gets infected by a particular disease and it's not resistant then the whole crop gets infected. And if that's the favorite wheat variety that's been grown in a country that year, you could potentially lose the whole of the country's crop. Avoiding such disasters is about ensuring resilience. And to do that you need diversity."

HELPING FUTURE CROPS THRIVE

Ruth is Kew's Project Coordinator for Adapting Agriculture to Climate Change, a ten-year project being run by Kew Gardens and the Global Crop Diversity Trust, which aims to address the issue of resilience. Specifically, the project aims to locate where the wild relatives of 29 popular modern crops grow, assess how much material from those species is already held in seed banks around the world, supplement existing collections with new seed stocks where gaps exist, and then make this genetic material available to "pre-breeders" (experts who identify desired traits in overlooked plants and begin to incorporate them into modern varieties) for selective breeding trials.

The long-term goal is to help breed some of the genetic diversity present in wild populations back into crop cultivars to make them more resilient to future climatic shifts. The 29 crops are among those covered under Annex I of the International Treaty on Plant Genetic Resources for Food and Agriculture (see page 22), and include barley, wheat, carrot, chickpea, potato, rice, sunflower, lentil, oat, banana, alfalfa, sorghum, rye, apple, common bean, and vetch.

GLOBAL RESEARCH

The crops chosen have, between them, around 450 crop wild relatives; that is, wild plant species that are closely genetically related to them. These encompass the ancestors of the domesticated crop plants and other closely related species. During the first three years of the project, scientists at Kew Gardens, the University of Birmingham, and the International Center for Tropical Agriculture (CIAT) worked to gather as much information on these plants as possible. They looked at where dried specimens of crop wild relatives held in herbaria around the world had originally been collected and combined this with habitat data to predict where the plants are likely to grow today.

They also investigated where seeds from crop wild relatives were already being stored for posterity in seed banks, and whether scientists knew how to germinate them. Using a technique called gap analysis, the scientists at CIAT modeled the distribution of living crop wild relatives and compared this to the information on seed bank collections. This enabled them to work out where gaps existed in the genetic resources of crop wild relatives available to farmers and pre-breeders.

The analysis showed that 54 percent of crop wild relatives on the list are of high priority, because either their seeds have not been collected before or the existing collections do not represent the full geographic distribution of the species. The countries with the highest number of priority crop wild relatives include Australia, Brazil, China, Cyprus, Ecuador, Ethiopia, India, Italy, Kenya, Mexico, Mozambique, Peru, Portugal, South Africa, Turkey, and the USA.

Although many of these countries lie within traditionally recognized centers of high wild crop diversity, the inclusion of the USA, Australia, and European countries on the list came as a surprise to the researchers. Unexpectedly, northern Australia proved to be rich in the wild relatives of the African staple crop sorghum.

COLLECTING WILD SEEDS

The MSBP is supporting each country in collecting seeds from its own crop wild relatives; duplicate seed collections will be stored in the country of origin and at the MSBP. Kew Gardens is providing national programs with training courses, technical advice and seed-collecting guides. The guides are designed to facilitate collection of high-quality and accurately identified seed batches. For each taxon, they provide identifying characteristics, photos including mature seeds, maps showing where collections should be targeted, and information indicating the best time of year to collect and the appropriate technique for collecting seeds. An ideal collection from one taxon includes a total of 20,000 seeds, comprising 20 collections from separate populations. However, collectors are advised to collect no more than 20 percent of seeds that are ready, so as not to affect the natural population. The seeds are cleaned in country and dried before being banked in cold stores. The MSBP will undertake germination tests on the seeds it receives to work out protocols that will awaken the seeds when needed.

Once seeds of all the crop wild relatives are safely stored, samples will be distributed to pre-breeders, who will experiment by growing up the crop wild relatives and investigating any traits that may prove to be useful within agriculture. "Crop wild relatives have just been sitting out there, evolving and dealing with whatever conditions have come their way," explains Ruth. "The pre-breeders have to grow the plants and see what they get. They try growing the plants under different conditions and then score them using set criteria. They might grow multiple batches, where they water some but not others, or they might deliberately infect some with a particular disease. Alternatively, they might expose them to extreme climates and look at what happens to each plant. Not all the traits are easy to predict, however, there's a cryptic diversity. For example, a small weedy wild rice plant would not be selected on its morphology for a breeding program aiming to increase yields. However, researchers have shown that, in some cases, offspring resulting from crosses of these plants with elite varieties produce increased yields."

NO TIME TO LOSE

Releasing a new crop variety to farmers can take 15 to 20 years. Given this long lead time, and the extent to which the effects of climate change are already influencing local weather patterns and cropping seasons, banking crop wild relatives needs to be done sooner rather than later. Pressures on the populations of wild species, from urbanization, climate shifts, and clearance for agriculture, mean some of the target populations for seed collections are already threatened. For example, a herbarium specimen of *Solanum ruvu* Vorontsova, a wild relative of the aubergine, was collected for the first time in Tanzania's Ruvu Forest Reserve in 2000. By the time it had been identified as a species that was new to science, the plant's native habitat had been destroyed. Botanists believe the plant is now extinct. Its demise represents a lost opportunity to make use of any favorable traits the plant may have carried.

HOW CROP WILD RELATIVES HAVE HELPED US BREED RESILIENT VARIETIES

Crop wild relatives represent a diverse and largely untapped source of genetic material that plant breeders have at their disposal to improve crops. They could potentially breed genes associated with drought-resistance into wheat, for instance. Some crops that have already been enhanced using the genetic variety within their crop wild relatives are highlighted here.

SALT-TOLERANT SUNFLOWERS

Domesticated 5,000 years ago, sunflowers (*Helianthus species*) now grow across 63 million acres (25 million hectares) in 70 countries. Their oily seeds are rich in protein, fat and linoleic acid (an important omega-6 fatty acid). Modern varieties have been created using genes from crop wild relatives to endow traits such as salt tolerance, healthier oil and resistance to pests and diseases. However, climate change is driving the need for new varieties able to thrive in hotter, dryer and even more saline conditions.

Salinity affects arid and semi-arid areas, when there is insufficient rainfall for continued percolation of water through the soil and where irrigated land has inadequate drainage. Growing wild in salt marshes, the pecos sunflower (*H. paradoxus*) can tolerate saline conditions better than the widely cultivated domesticated sunflower *H. annuus*. Salt-tolerant hybrids developed using *H. paradoxus* promise a 25 percent higher yield when grown in salt-rich soils. However, the pecos sunflower is a threatened species, growing only in six small areas of Texas and New Mexico, USA.

ROSY-RED TOMATOES

The impacts of crossing cultivated crop varieties with their wild relatives are not always predictable. For example, genes from *Solanum habrochaites*, a wild relative of the cultivated tomato, have been used to improve fruit color even though the fruits of *S. habrochaites* do not turn red when ripe. Scientists found that including *S. habrochaites* traits in breeding lines improved fruit color by 33 percent. Meanwhile, yield increased by 48 percent and the soluble solids content by 22 percent.

DISEASE-BEATING BROCCOLI

Eating broccoli is associated with a reduced risk for several cancers. This is because it contains the compound glucoraphanin, which forms sulforaphane when the vegetable is chewed and digested. Studies suggest that sulforaphane inhibits cancer by increasing the production of enzymes that detoxify carcinogens and inhibiting other enzymes that activate carcinogens.

Broccoli is a cultivar of the species *Brassica oleracea*, as are cabbage, kale, Brussels sprouts, and cauliflower. *Brassica villosa* is a wild relative of broccoli, from Sicily and southern Italy, which is rich in glucoraphanin. By crossing the two, scientists at the John Innes Centre and the Institute of Food Research (in Norwich, UK) created a glucoraphanin-rich variety of broccoli named "Beneforté." Launched in 2011, it is now available in many supermarkets.

Research conducted in 2013 suggests that sulforaphane may also prevent or slow osteoarthritis.

HIGH-YIELDING RICE

Three out of four of the world's most populous nations—China, India, and Indonesia—eat rice-based diets. With populations rising in these countries, increasing rice yields is highly desirable. All commercial varieties of rice derive from *Oryza sativa*. In the 1990s, scientists at Cornell University, New York, USA, began experimenting with crossing a dozen *O. sativa* varieties of rice with their wild relatives *O. rufipogon*, *O. glaberrina*, and *O. barthiiaro*.

Not only did the experiments result in up to 20 percent higher yields, but one cross proved to be resistant to the Hoja blanca virus, the most destructive viral disease affecting rice in Latin America. This was despite the fact that neither of the parent crops was resistant. Earlier work conducted in the 1970s by the International Rice Research Institute had bred resistance to grassy stunt virus, a disease affecting crops in southeast Asia, into crops using genes from the wild rice species *Oryza nivara*. Varieties carrying resistance from *O. nivara* are still grown by farmers in the region today.

CLIMATE-RESILIENT WHEAT

With climate change threatening to increase droughts and raise salinity levels in certain parts of the world, the race is on to find crop varieties that can tolerate drier and saltier conditions. Wheat is a highly important crop, providing 21 percent of food calories globally.

Emmer wheat (*Triticum dicoccoides*) is a wild parent that has already contributed some genes to domesticated wheats. *T. dicoccoides* has been found to include both salt- and drought-tolerant genes. This genetic material has great potential for helping to make modern wheats more resistant to the effects of climate change.

HUMAN USES OF SEEDS DOWN THE AGES

PRE-3.5 MA

Chimp-like human ancestors primarily ate leaves and fruit from trees, shrubs, and herbs.

3.5 MA

Early hominids such as *Australopithecus afarensis* and *Kenyanthropus platyops* moved out of forested areas and onto the plains of eastern Africa. With this change of habitat came a dietary shift to eating more grasses, sedges, and succulents.

2.2 MA

The first true human, *Homo habilis*, appeared with a larger brain and smaller teeth than its primate ancestors. Its range was limited to eastern and southern Africa.

1.89 MA

Homo erectus evolved, exhibiting similar body proportions to modern humans. This species' adaptations suggest it lived entirely on the ground and was able to walk or run long distances.

790,000 YEARS AGO

The first evidence of controlled use of fires by humans comes from archaeological remains dated to this time at Gesher Benot Ya'aqov in Israel.

60,000 YEARS AGO

The last ice age began.

50,000 YEARS AGO

Phytoliths (plant remains preserved as silica) and starch grains found in the teeth of remains from inhabitants of Shanidar Cave, northern Iraq show that Neanderthals were eating grass seeds (Triticeae cf. *Hordeum*), dates (*Phoenix*), legumes (Fabaceae), and plant roots and tubers at this time. Cooking was also evident. The macrobotanical evidence found in the cave matches well with other Neanderthal remains in western Asia.

12,000 YEARS AGO

The transition from hunting and gathering to agriculture and domestication began, possibly prompted by changes in weather patterns.

10,000 YEARS AGO

The last ice age drew to a close. The warmer, wetter weather helped promote growth of grasslands and forests. The first fully domesticated crops—including einkorn and emmer wheat, lentils, and peas—grew in west Asian villages at this time.

8,000 YEARS AGO

Maize was first cultivated in Mesoamerica; this was the staple that underpinned the Incan, Mayan, and Aztec civilizations.

7,000 YEARS AGO

The earliest cereal cultivation took place in North Africa. The earliest evidence of rice cultivation in China is dated to this time.

5,000 YEARS AGO

Potters were creating wheel-shaped ceramic pots for storing food. Sumerian civilization emerged in southern Mesopotamia (modern-day southern Iraq); Sumerian farmers used some of the earliest large-scale irrigation methods to cultivate grains and legumes. They used 40 percent of their harvest to produce 19 types of beer.

4,000 YEARS AGO

Frescos from the tomb of Iti, El Gebelein, Egypt, dated to this time, show workers transporting grain to granaries for storage.

3,337 YEARS AGO

The Egyptian boy pharaoh Tutankhamun died at the age of 18. He was buried along with an array of plant material, including emmer wheat, chickpea, barley, and watermelon seeds.

2,000 YEARS AGO

The Romans clearly understood the importance of seed selection for high-yielding agriculture. The Roman scholar and writer Marcus Terrentius Varro wrote, "Ears of the finest and best crop should be taken to the threshing-floor and kept separate from the rest, so that the farmer may have the best possible seed." Greco-Roman communities made necklaces out of seeds, fruit, leaves, and flowers at this time.

1,500 YEARS AGO

By the Middle Ages, seeds of the pepper plant (*Piper nigrum*) had become very important in Europe as a food flavoring.

1492

Christopher Columbus became the first European to encounter the Americas. At the time, he was seeking the revered "Spice Islands" of the East by journeying west across the Atlantic. His "discovery" prompted the "Columbian exchange," in which plants, peoples, animals, and knowledge were subsequently exchanged between the New World and the Old. Plants introduced to the Old World included maize, groundnut (peanut), pumpkin, and many beans.

1498

Also motivated by finding supplies of spices, Vasco da Gama became the first person to sail from Europe to India. His journey paved the way for colonialism.

1514

Nutmeg and mace, derived from the nutmeg fruit (*Myristica fragrans*), had long been popular in Europe. When the Portuguese reached the Moluccas (the "Spice Islands" of the Malay archipelago) in 1514, they gained a monopoly in the trade.

1550s

In Antwerp, the price of pepper acted as a barometer for European business.

EARLY 1800s

British farmers celebrated finishing their sowing with a seed festival, at which a currant or plum "seed cake" was served. Today, seed cake refers to a cake flavored with caraway seeds, a kind which was once popular in Britain.

1876

Rubber seeds "stolen" by Henry Wickham from the Brazilian Amazon at the request of the British Government, and grown into seedlings at Kew Gardens in London, were sent to Ceylon (now Sri Lanka) to be a potential British-controlled source of rubber. Descendants of these plants went on to found today's global rubber industry.

1994

The Flavr Savr tomato was the first genetically modified crop to be sold in the USA. It ripened more slowly than traditional varieties, giving it a longer shelf life.

1998

Mustard oil contaminated with toxic poppy seeds (*Argemone mexicana*) killed more than 50 people and made thousands seriously ill in northern India.

2000

Prince Charles opened Kew Gardens' MSBP, describing it as "a gold reserve...a place where this reserve currency, in this case life itself, is stored."

2001

The headless torso of a boy, named by police as "Adam," was pulled from the River Thames in London. Scientists at Kew Gardens later found that he had been poisoned with the calabar bean (*Physostigma venenosum*), a plant that has been linked to African witchcraft practices.

2002

Seed company Pioneer Hi-Bred International flew soybean seeds to the International Space Station. When scientists later analyzed seeds from soybean plants grown from the space seeds, they found slightly more carbohydrate and a little less oil than in the control seeds. But these figures were all within the normal range.

2009

Scientists at the University of Maryland, USA, designed a mini-helicopter with a single blade, emulating the way maple seeds spin in flight. The researchers filmed seeds with a high-speed camera and then copied their design.

2014

Scientists at Rothamsted Research modified seeds from *Camelina sativa* (false flax) to contain omega 3 fatty acids normally found in oily fish. They did so by modifying the seeds with genes from microalgae, which naturally produce the fatty acids. The oil was incorporated into salmon feed to assess its suitability as an alternative to wild fish oils. In nature, fish do not produce omega-3 oils; instead they get them from algae within the marine environment. Farmed salmon have to be given feed containing fish oil, so that it provides the essential fatty acids desired by consumers.

SEEDS WORTH THEIR WEIGHT IN DIAMONDS

The term "carat" used to measure the weight of a diamond derives from the name of the carob plant (*Ceratonia siliqua*). Carob seeds were used by Mediterranean traders as a unit of measurement; a jewel that weighed the same as five seeds became known as five carobs, or five carat, in weight. Although carob seeds do not all weigh the same, a study conducted in 2006 found that people can discriminate differences in carob seed weight of around five percent by eye. Therefore merchants would have been able to select seeds of approximately the same weight for their scales. The average weight of a carob seed was found to be 0.197 grams (about $^{1}/_{150}$th of an ounce); this was standardized to 200 milligrams in 1907. This figure is still used today; there are five carats per gram.

THE FATHER OF SEED SCIENCE

1 *Nikolai Vavilov, who dedicated his life to collecting wild crop seeds. He hoped botanists might use the genetic diversity they contained to breed disease-resistant cultivars, and prevent famines.*

2 *Vavilov's notebooks provide a valuable record of his research.*

Among the first people to realize the importance of genetic diversity in crop plants was the Russian botanist, geneticist, and plant breeder, Nikolai Vavilov. Born in 1887, Vavilov grew up in a village where crop failures frequently caused hunger and hardship. At a time when most botanists were concerned with wild plants, Vavilov instead began classifying and stockpiling the genetic diversity contained within cultivated plants. He believed that the emerging science of genetics was crucial: it would enable botanists to use the diversity that exists within seeds to develop new disease-resistant crop cultivars that would help prevent starvation. "However rich nature may be in forms, the combinations of characters that would perfectly suit man would be extremely rare, and the deliberate creation of new and agriculturally more advanced forms constitutes a current objective of plant science," he wrote.

Early in his career, Vavilov worked at the Moscow Agricultural Institute and also spent time in the UK at the John Innes Horticultural Institute at Cambridge University. His interests lay in plant breeding, disease resistance, and genetics. He eventually took up the post of Head of the Department of Applied Botany and Plant Breeding in Petrograd (now St. Petersburg). "I would like the Department to be a necessary institution, as useful to everybody as possible," he explained in a letter in 1920. "I'd like to gather the varietal diversity from all over the world, bring it to order, turn the Department into the treasury of all crops and other floras, and launch the publishing of 'Flora Culta,' the botanical and geographical study of all cultivated plants. The outcome is uncertain, especially considering the surrounding hunger and cold. But still, I want to try."

AGRICULTURE'S CENTERS OF ORIGIN

Over the course of his career, Vavilov undertook more than 115 research expeditions to 64 countries, including Ethiopia, Italy, Kazakhstan, Mexico, Brazil, and the USA. He deliberately sought out the areas in which agriculture originated, with the specific aim of finding useful genes for modern needs, both in wild relatives of crops and in traditional landraces grown by farmers. He used the knowledge he gained on his explorations to develop a theory on the "center of origins." He believed that each type of cultivated plant originated from a particular place where the greatest variation in that crop could still be found. He called these places centers of origin. In 1926 he published the paper "The centers of origin of cultivated plants," identifying five major foci of the main field, garden, and orchard crops. These were primarily located between 20° and 40° N latitude within major mountain belts.

By then, with the support of Premier Vladimir Lenin, Vavilov had turned his Department into a vast plant-breeding empire. At the heart of the Department lay thousands of live seed and tuber samples that he and his colleagues had collected on their travels. Vavilov's contribution to agricultural science was not appreciated by Lenin's successor, Stalin, however, who rose to power in the late 1920s. In August 1940, while Vavilov was gathering wild grass specimens in the Carpathian Mountains, four men in a black sedan arrived to say he was urgently needed in Moscow. In fact, they had come to take him to prison in Saratov.

SEEDS UNDER SIEGE

While Vavilov was incarcerated, and Germany was gaining ground across Europe in World War II, Stalin began shifting half a million treasures, including paintings, frescos, and gems, from Russia's famous gallery, the Hermitage, to secret hiding places to protect them from Hitler's advancing armies. He felt no need to safeguard the thousands of seeds, roots, and fruits of the 2,500 species of food crops that were by now residing close by in Vavilov's seed bank. But the staff were not prepared to let Hitler take the immensely valuable resource that they and Vavilov had painstakingly assembled. With Leningrad (as Petrograd had become) now cut off, and its people starving, they barricaded themselves inside the building and took it in turns to watch over the seed bank. In the end, nine of Vavilov's co-workers either starved to death or died of disease, refusing to eat the seeds that could have saved them.

VAVILOV'S LEGACY

Meanwhile Vavilov, who had roamed five continents to gather wild crop seeds to prevent humanity suffering from the famine and hunger he had witnessed as a child, also slowly succumbed to starvation, dying in prison in 1943. Only after Vavilov's death did the value of his pioneering work emerge. According to the Russian food historian G.A. Golubev, writing in 1979, "Four-fifths of all the Soviet Union's cultivated areas are sown with varieties of different plants derived from the seeds available in the VIR's [abbreviation for N. I. Vavilov Institute of Plant Genetic Resources, which the Department is now called] unique world collection." Meanwhile, his map of the center of origins of food crops compares well with the modern maps that scientists at Kew Gardens' MSBP have made in their search for crop wild relatives. "Looking at our distribution map, it's amazing that Vavilov's insight all that time ago, with very limited data compared to what we have now and limited tools to analyze it, was actually pretty accurate," says Ruth Eastwood.

2

SEED BANK RUSSIA

THE SEED BANK THAT SURVIVED A SIEGE

1 *The grand building that houses the N. I. Vavilov Institute of Plant Genetic Resources.*

2

КОЛЛЕКЦИЯ ПШЕНИЦ № 12730
Отдел прикл. бот. Г.И.О.А.
Tr. compactum Host. icterinum Al. (примесь)
Поездка Н.И.Вавилова по Афганистану — VII-XII 1924 г.
В 15 верстах к Ю-З. от Кабула в Кандагарском направлении
Собр. Н. И. Вавилов 17/IX-24г.

3

2 *Vavilov's portrait keeps a watchful eye over the latest generation of seed researchers.*

3 *Labels on each herbarium specimen provide valuable information.*

NAME

N. I. Vavilov Institute of Plant Genetic Resources (VIR), St. Petersburg, Russia

NUMBER OF ACCESSIONS

380,000 gene types representing 65 plant families, 380 genera, and 2,500 plant species.

WHEN FOUNDED

The Bureau of Applied Botany was formed in 1894 under the auspices of the Scientific Committee of the Ministry of Land Cultivation and State Property. The remit of its scientists was to study the taxonomy of wild and cultivated plant species. In 1916, the Bureau was renamed the Department of Applied Botany and Plant Breeding. Under Nikolai Vavilov's directorship from 1920, the Department was reorganized and became the Institute of Plant Industry. Its current name reflects Vavilov's acclaimed contribution to the institution and to seed science.

FOCUS OF THE COLLECTION

The collection is focused on major crops, including wheat (54,000 accessions), rye (3,000 accessions), barley (20,000 accessions), oats (13,000 accessions), legumes (42,000 accessions), small grains such as sorghum (57,000 accessions), plants yielding oil and fiber (20,000 accessions), forage plants (20,000 accessions), potatoes (10,000 accessions), vegetables and melons (49,000 accessions), and fruits (23,000 accessions).

WHY IT IS NEEDED

Research into crop plants is more important than ever. The world's population is forecast to rise to 10.6 billion by 2050. Governments must find a way to feed those people at a time when resources of land, soil, and water, and ecosystem services such as pollination, are under increasing pressure.

WHO FUNDS IT

The Russian Government funds VIR's activities through the Federal Agency of Scientific Organization. However, this financial support is 30 times less than the budget of the National Plant Germplasm System of the United States Department of Agriculture's Agricultural Research Service.

WHERE SEEDS ARE STORED

Collections are held at the headquarters in St. Petersburg, as well as at 12 experimental stations located across European and Asiatic Russia. The National Seed Store was constructed at Kuban Experiment Station of VIR (in the Krasnodar region of the Russian Federation) in 1976. The seed store is a three-story building with the first floor below ground level. The storage chambers are located on the ground and the first floors; there are 12 on each floor. In these chambers, VIR has more than 300,000 accessions stored at 39°F (4°C). Today, there are additional stores kept at 39°F (4°C), 14°F (-10°C) and -4°F (-20°C), as well as liquid nitrogen facilities, in the main building of VIR in St. Petersburg. In total, VIR has 550,000 accessions of cultivated species and wild relatives held at the lower temperatures.

1 *Boxes of seeds line the walls; in total VIR holds 550,000 accessions.*

2 *Specimens of dried plants can help to identify species and cultivars.*

3 *Cryopreservation is used to store seeds that cannot be dried.*

4 *A scientist processes a batch of seeds ready for storage.*

CURRENT RESEARCH

All the Institute's activities adhere to Vavilov's principles of applying an integrated approach to collecting, studying, preserving, and using plant genetic resources. Research on comparative genetics of the major species of cultivated plants and their wild relatives is combined with solving evolutionary, taxonomic, and breeding problems, as well as meeting the demands of agricultural production.

SEEDS WITH A STORY

The majority of VIR's collection exists nowhere else; while other nations have lost duplicate material, VIR's collections have endured despite economic difficulties and war, specifically the Siege of Leningrad. For example, all the seed material that Vavilov collected in Ethiopia in 1927 is unique. Emperor Haile Selassie had met Vavilov that year, while still the Regent Ras Tafari. He wrote an official letter to the Central Committee of the Communist Party of the USSR, asking if the seed material Vavilov had collected in Ethiopia could be sent back. When the letter was passed to the Institute, its specialists prepared duplicates of the accessions and presented them to the Ethiopian people. In the 1980s, Ethiopia and Eritrea suffered debilitating droughts and fought a civil war. As a result, most of the two nations' landrace crop diversity was lost. However, the genetic material in seeds collected by Vavilov endures to this day.

PLANTS AND SEEDS FROM THE WORLD'S ARID LANDS

Around 40 percent of the Earth's surface can be classified as drylands, where the average rainfall is less than the potential moisture losses through evaporation and transpiration. These areas support two billion people, 90 percent of whom live in developing countries. Plants that can tolerate the extreme conditions are often of great importance as foods, fodder, medicines, dyes, and building materials. The fact that the plants favor dry conditions is usually beneficial when banking their seeds, as drying is an integral part of the storage process. Here are some species that are stored for posterity in Kew Gardens' MSBP seed vault.

ACACIA SENEGAL

This drought-tolerant bush grows in thickets in many African countries. It thrives in areas with rainfall limited to between 8 and 24 inches (20 and 60 cm) a year, including those on the edge of the Saharan and West Asian deserts. It can endure up to 11 months of drought and temperatures of 113°F (45°C), sending out deep roots to soak up water from far below the surface.

Acacia senegal is an important economic plant, supplying 90 percent of the commodity gum Arabic. Produced from the hardened sap of the acacia, this gum is used as an emulsifier and stabilizer in fizzy drinks, sweets, and medicinal lozenges. Sudan is the largest producer of the commodity, with millions of people there depending at least in part on gum Arabic for their incomes.

This botanical illustration depicts Acacia senegal, *the ubiquitous African plant that yields gum Arabic, an emulsifier and stabilizer used within the food and pharmaceutical industries.*

Anogeissus leiocarpa has fragrant yellow flowers. Its leaves are used to dye Mali's distinctive mud cloth yellow.

The flat, brown seeds of *A. senegal* are sometimes eaten as a vegetable. Unless freshly shed, they have "physical dormancy" (see Chapter 5), which means they have an impermeable coating that must be broken to allow water in before they will germinate. For cultivation purposes, this is achieved by chipping the coat with a scalpel or soaking the seeds in sulphuric acid.

MORINGA PERIGRINA

The fast-growing but endangered yusor tree, as *Moringa peregrina* is known locally, lives at the base of the remote mountains of Jordan's southern Rift Valley. Bedouin communities eat its seeds either raw, fried, or roasted, as well as boiling them to produce an oil. Its wood is a source of firewood, while the leaves provide fodder for goats. Seeds of the tree gathered in 2005 are now preserved at the MSBP and at Jordan's National Center for Agricultural Research and Extension (NCARE). In June 2014, a plant of *Moringa perigrina* flowered at Kew Gardens, the first time the species had bloomed outside its native habitat.

Grazing animals have had a devastating impact on Jordan's natural vegetation, so the search is on to find plants that can withstand constant nibbling. NCARE has conducted trials of several native dryland plants that can tolerate grazing, including *Moringa peregrina*. Results suggest that if regularly pruned, the yusor tree could provide a permanent source of green fodder for livestock. If propagated on a large scale, this could help to relieve pressure on natural ecosystems.

ANOGEISSUS LEIOCARPA

Anogeissus leiocarpa is a graceful tree that grows across tropical Africa's savannas. It is used to make chewing sticks and to treat worms but is best known as a dye used in Mali for producing *bògòlanfini* or mud cloth. When cotton cloth is soaked in water with leaves from *Anogeissus leiocarpa*, it turns yellow. Fermented mud is then used to paint patterns on the cloth. Large quantities of leaves are needed to meet the demand for mud cloth from local and export markets.

Historically, leaves of *Anogeissus leiocarpa* were harvested from wild populations, but Mali and Burkina Faso have both now initiated commercial cultivation of the tree. The MSBP's Useful Plants Project helps communities learn how to germinate seeds of beneficial native plants in order to conserve wild populations. Seedlings of *Anogeissus leiocarpa* are among the young plants now thriving in a local community nursery, as a result of the project.

ACANTHOSICYOS HORRIDUS

The generic name of this desert plant comes from the Greek words *akantha*, meaning thorn, and *sikuos*, which translates as cucumber. *Horridus*, meanwhile, from the Latin verb *horrere*, refers to the sharp spines that grow on its arching, leafless branches. It can tolerate extreme drought, growing on Namibia's parched Sossusvlei sand dunes, as well as in neighboring Angola. It usually inhabits locations close to rivers, where its taproots grow down as far as 130 feet (40 m) to suck up water. Its grooved stems trap sand and help to stabilize sand hummocks.

Acanthosicyos horridus is also known as the Nara melon on account of the large, spiky yellow fruits it produces. Namibia's Topnaar people leave the fruits in the sun or bury them in the sand so that they soften. They then take off the peel and boil the enclosed pulp to remove the seeds. These are dried in the sun and eaten like nuts; they taste a little like almonds. Archaeologists have found remains of seed shells from the Nara melon dating back 8,000 years, indicating that the plant has been used by local people for millennia.

HARPAGOPHYTUM PROCUMBENS

This perennial herb of the sesame seed family (Pedaliaceae) grows amid the Kalahari sands of southern Africa in Angola, Botswana, Zambia, Zimbabwe, Namibia, Mozambique, and South Africa. Its common name of devil's claws refers to its spiny fruits; the enclosed seeds are dispersed by animals when the fruit gets stuck on their claws or feet and they trample it, thus releasing seeds. The devil's claw has been used medicinally by the Kalahari people for centuries to aid everything from digestive disorders to skin infections. It is now also widely used in Western medicine. According to the University of Maryland Medical Center, several studies have found that taking devil's claw for two to three months reduces pain and improves movement in people with osteoarthritis. It is also used in veterinary medicine.

Wild harvesting of *Harpagophytum procumbens* is putting the species under pressure. In 2003, a propagation study conducted by the MSBP and the Botswana College of Agriculture sought to assess the potential for cultivating eight important plant species, including the devil's claw. A method for propagating *Harpagophytum procumbens* without fertilizer or water was developed for use by small-scale farmers in dry regions. According to the Food and Agriculture Organization of the United Nations (FAO), the projected export value of products from *Harpagophytum procumbens* in Namibia alone is $2.7 million. However, in Namibia there is limited cultivation because of concern that this would be to the detriment of communities that sustainably harvest the plant from the wild. The trend in Namibia is therefore toward "enrichment planting" or rehabilitation of unsustainably harvested areas rather than traditional cultivation.

SEEDS AT TWO ENDS OF A SCALE

Seed banking experts recognize two main types of seeds: orthodox ones, which can tolerate drying, and recalcitrant ones, which cannot. Only seeds that can be dried can be stored for any length of time. This is because they are kept at around -4°F (-20°C) to slow down their metabolism. At such low temperatures, remaining water turns to ice, rupturing the seed's cells and preventing germination. Fortunately, between 70 and 80 percent of seeds can tolerate drying and can therefore be stored.

The remaining 20 to 30 percent are recalcitrant and thus harder to bank. Cryopreservation techniques are now being developed, with one method involving the use of liquid nitrogen to rapidly cool the plant embryo to below the freezing point of water.

A seed's type relates to its natural habitat; orthodox seeds tend to come from more arid areas, while recalcitrant ones live in more humid climes. Intermediate seeds fall somewhere on a scale between orthodox and recalcitrant.

1 *Kew scientists collected seeds of* Acanthosicyos horridus *from the Swakopmund river valley in Namibia.*

2 *The plant's vicious spines make gathering its seeds quite a challenge.*

3 *The spiny fruit of* Harpagophytum procumbens *give rise to its common name of devil's claw.*

GRASS PEA
(LATHYRUS SATIVUS)

GENUS Lathyrus

FAMILY Fabaceae

SEED SIZE ¼ inch (8–9mm)

TYPE OF DISPERSAL Ancient wild grass peas had pods that split to shed their peas; a single mutation of a gene preventing this 'dehiscence' aided domestication of grass pea crops by early farmers

SEED STORAGE TYPE Orthodox. Viability is halved after 5 years of open storage at room temperature

COMPOSITION Oil: 0.76%; Protein: 31.01%

With climate change forecast to cause more droughts and floods, finding crops that are hardy enough to withstand extreme weather events is a priority. One plant with good credentials for this role is the grass pea (*Lathyrus sativus*), so named because of its grasslike resemblance. Extremely hardy and easy to cultivate, it bears seeds that are tasty and contain good quantities of nutritious protein. The only problem is, they are also toxic. When people have eaten the seeds as a major part of their diet for a prolonged period, such as during famines, they have suffered from neurolathyrism, a disease that permanently paralyzes adults from the knees down and causes children to suffer brain damage.

In fact, the potent neurotoxin that renders it poisonous when eaten in large quantities also makes the plant hardy in extreme conditions. When a crop experiences a high level of water stress, such as during a prolonged drought, the levels of neurotoxin in the plants increase. This means that the times when people are most likely to eat grass peas coincide with when they are most poisonous to humans. New cases of neurolathyrism have emerged following each major famine and food shortage in Ethiopia since the mid-1970s. To give one example of its devastating impact on a community, an epidemic in the 1970s in the Gondar region of the country permanently crippled one percent of the population.

Wild populations of grass pea are genetically diverse, with African and Asian grass pea plants being seven times more toxic than Middle Eastern ones. Breeders at the International Center for Agricultural Research for Dry Areas, in Syria, and the Centre for Legumes in Mediterranean Agriculture, in Australia, have managed to create hybrids with sufficient levels of neurotoxin to maintain their hardiness but within a range that makes them safe to eat. The plant is one of the 29 key species being collected and studied as part of the Adapting Agriculture to Climate Change project, which will facilitate future work. The hope is that rural communities may soon be able to rely on the grass pea as a famine food, without fearing the consequences of eating it.

In times of famine, some African communities have little choice but to eat the toxic seeds of the grass pea, Lathyrus sativus, *which can cause paralysis.*

CHAPTER 2

HOW PLANTS EVOLVED ON PLANET EARTH

TINY ALGAE GIVE RISE TO THE FIRST PLANTS

There are around 400,000 species of plants on Earth. Each one, from the tiniest mosses to towering hardwood trees, from the plants we eat to those that are deadly poisonous, has evolved over millions of years, adapting to the environmental conditions around it to maximize its chances of survival. This is why the banana has large leaves that can catch the light of shaded tropical rainforests; conifers growing at high altitudes have sloping branches that easily shed accumulated snow; and succulents have fleshy leaves that store water they can draw on in arid deserts.

Some ancient plant lineages changed little after their emergence and still exist in their original form today; others yielded new evolutionary paths that gave rise to thousands of different plant species. And along the way, many unsuccessful plants became extinct, leaving rare fossils and genetic clues as the only evidence of their existence. Botanists study both living plants and fossils to untangle the convoluted evolutionary path plants have taken to create the world's modern-day flora.

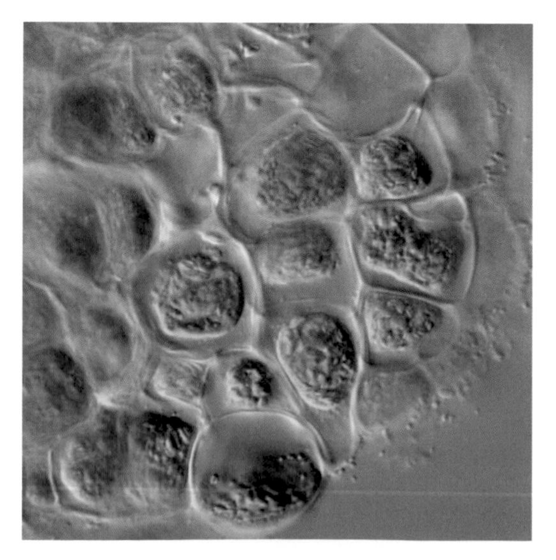

Coleochaete, *a genus of green algae that grows in disc-like colonies. Classified within the Charophyta division along with Charophycean algae,* coleochaetes *are among the closest living relatives of land plants.*

Research undertaken so far tells us that all land plants originated from Charophycean algae. Scientists arrived at this conclusion after studying how the cells of these algae divide and also their chemical characteristics. Algae originated in the ocean but Charophycean algae inhabit freshwater environments. When they made the move from the ocean to inhabit small bodies of water on land some 450 million years ago, they set in motion an evolutionary process that resulted in the great diversity of plants that currently live on Earth. We know that modern algae require water to reproduce; they release spores into water currents. So the ancient Charophycean algae living in small water bodies would have been at risk of desiccation, a pressure that inevitably drove a selection process for plants that could survive temporary periods of dryness. Those organisms that evolved to produce desiccation-resistant spores and a water-resistant layer that retained moisture soon gained an upper hand. And so the evolution from algae to plants began.

SUN SEEKERS

Algae were not the only organisms trying their luck at living on land at this time; fungi and lichens would have grown alongside them in freshwater pools. Algae, and the early land plants they gave rise to, needed light to survive, so the next evolutionary driver was the ability to reach for the sun. Those plants that gained in stature reached more sunlight than those organisms growing beneath them and in doing so they gained another benefit: more efficient spore dispersal. "If you produce your spores a little bit above the ground, they're going

Illustration depicting Charophycean algae. Research indicates that these algae gave rise to all land plants.

to disperse much better than if you produce them at ground level," explains Peter Crane, Dean of the School of Forestry and Environmental Studies at Yale University, USA. "You get them up into the moving air and those spores can then drift around much more easily. And as you get bigger you then need structural support. So then you start evolving water-conducting tissues to bring water up from the base of the plant to the aerial parts of the plant. Then, when you really get bigger, you need something other than your internal hydrostatic skeleton to hold you up."

HOW LIFE EVOLVED ON EARTH

Scientists believe that the earliest life forms were simple single-celled microbes that lived some four billion years ago around deep-sea hydrothermal vents. These are rock chimneys, formed where the Earth's crust spreads apart, enabling hot (660–750°F / 349–399°C) fluids, minerals, and gases to rise up from the planet's depths. These ancient vents were rich in the requirements of life: hydrogen, carbon dioxide, and minerals containing iron, nickel, and sulphur. Single-celled prokaryotic organisms fueled by sulphur from the vents are thought to have kick-started life on planet Earth.

Over time, these simple organisms gave rise to eukaryotes, typically multi-celled organisms that include plants, animals, and fungi along with many protista (which include species of red, green, and brown algae). Scientists know that chloroplasts, the photosynthesizing organelles in plants that convert sunlight into food energy, are derived from cyanobacteria and that mitochondria, the power-houses of cells found in animals, formed from purple bacteria. They believe that these two types of bacteria were enveloped into larger cells between 1,600 and 2,700 million years ago, resulting in their presence in plants and humans today.

In 2013, scientists from the National Institute for Basic Biology in Okazaki, Japan, suggested that the evolutionary advance from prokaryote to eukaryote may have come about through the algae eating the bacteria. They drew their conclusions after studying the single-celled *Cymbomonas*, one of the oldest extant groups of algae, which usually derive their energy from photosynthesis but under conditions of low light were seen to switch to eating bacteria. This they did by sucking the bacteria up a tube into a vacuole, a form of tiny stomach where digestion takes place. It is possible that certain bacteria sucked up into ancient algae may have "escaped" by breaking through the feeding tube or out of the vacuole. These bacteria would then have continued to live inside the algae and their plant descendants thereafter.

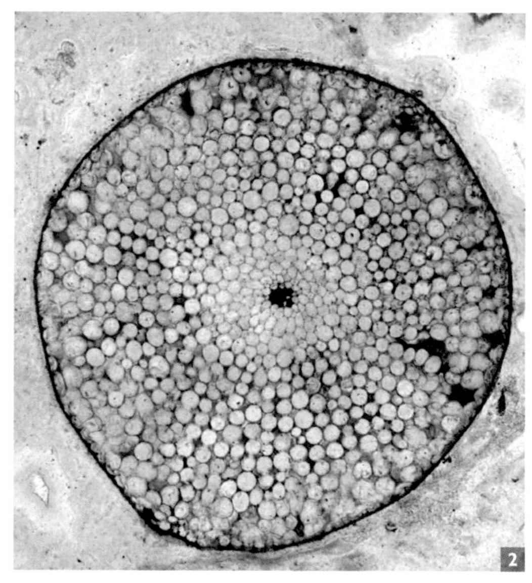

1 *The first vascular plants, known as tracheophytes, were the swamp-dwelling* Cooksonia.

2 *A thin section of Rhynie chert, a silica-rich rock found in Scotland, showing the cross-section of a stem of the early land plant* Rhynia gwynne-vaughanii.

FOSSILS HOLD CLUES TO EARLIEST LAND PLANTS

The earliest potential fossil evidence for land plants comes from bryophytes—plants related to modern mosses, liverworts, and hornworts. The evidence, which includes spores, sheets of cells, and tubes related to a form of non-supporting (nonvascular) water-transport system, can be seen in remnants found in deposits dated to around 470 Ma in Argentina, the Czech Republic, and Saudi Arabia. However the earliest convincing evidence for bryophytes—in the form of tiny structures, known as sporangia, in which spores develop—was found in Oman and dates to around 450 Ma. Later, around 440 Ma in the early Silurian period, the fossil record indicates a shift from spores which appear to be nonvascular to vascular spores (having lignified tissues, called xylem, with which to transport nutrients and water around the plant). Based on this evidence, scientists believe the advance to vascular plants occurred at this time, the vascular system answering the need for a frame to support plants as they grew bigger, while enabling them to circulate water and nutrients to all their parts. From now on plants were all set to grow much taller.

Fossils of the first vascular plants, *Cooksonia*, were first described in 1937 and were named after Isabel Cookson, an Australian botanist who had collected fossil specimens from Wales some three years earlier. These leafless, mosslike plants grew to around 2½ inches (6-6.5cm) high, and had stems that split into two and were topped with sporangia containing spores. Living in swamps, they were seemingly widespread across Europe and North America, with the earliest fossils found in Ireland and dating to 425 Ma.

Other early land plants include *Aglaophyton major* and *Rhynia gwynne-vaughanii*, fossils of which are preserved in Scotland's Rhynie chert. This sequence of sedimentary rocks, composed of silica minerals, contains fossils of an entire suite of land and freshwater plants, animals, bacteria, and fungi. By 420 Ma, vegetation of a variety of small vascular plants existed on many continents.

TWO-PRONGED APPROACH TO EVOLUTION STUDIES

Botanists have two approaches when it comes to trying to unravel the evolution of plants. They study living plants using the concept that "the present is the key to the past," and they examine fossil plants following the idea that "the past is the key to the present."

Looking at living plants reveals similarities in features such as flowers, seeds, and the way nutrients are carried around plant superstructures. This has enabled botanists to construct a classification of plants, which they rearrange as and when new information becomes available. The emergence of molecular techniques since the early 1990s has extended this capability to plants' genetic makeup (DNA), so that researchers can now make classifications according to the structure of an organism's DNA.

Delving into the fossil record, meanwhile, enables scientists to obtain information on plants that once existed but are now extinct, as well as dating particular evolutionary shifts. This kind of analysis has helped fill out the "evolutionary tree" of plant life. "Adding in information from the fossil record can be particularly helpful in revealing 'missing links,'" says Peter Crane, Dean of the School of Forestry and Environmental Studies at Yale University, USA.

"For example, among living plants you have 100 percent correlation between all plants that produce seeds and all those that produce secondary xylem [rigid, lignified tissue] and have the ability to form trees. When you go back to the fossil record you find there's a whole group that had secondary xylem and the ability to form trees but which had not yet evolved seeds. So they're halfway to being modern seed plants.

There's a similar story with the fossil bird *Archaeopteryx*. It's a bird in the sense that it has feathers and various skeletal features akin to modern birds but it hasn't lost all its teeth, which birds don't have. So you have this curious intermediate thing that's well on the way to being a bird but it's still hanging on to its reptilian teeth."

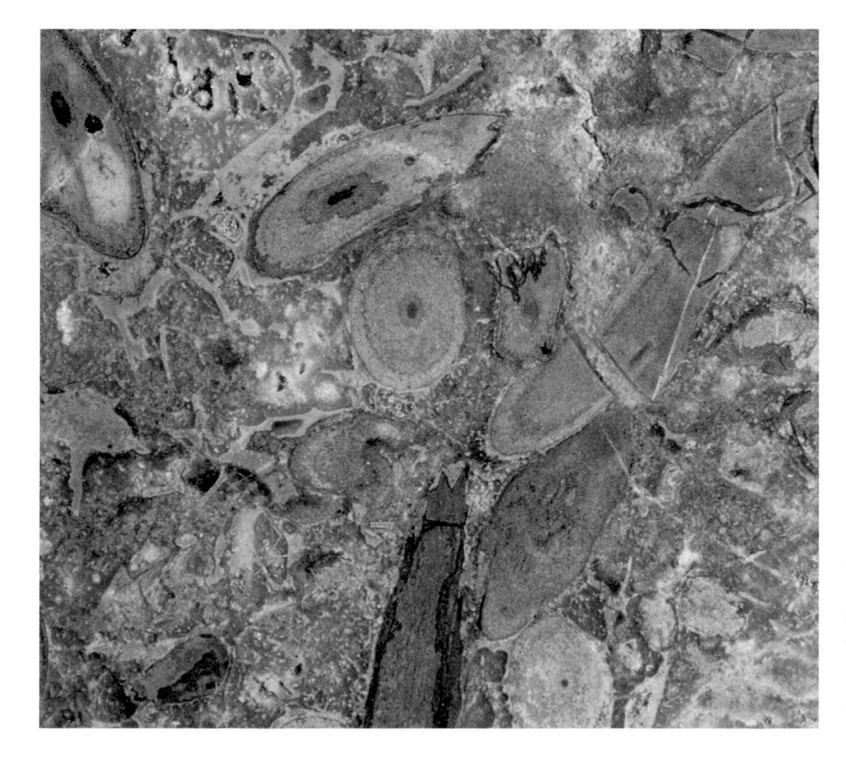

A thin section of Rhynie chert showing the stems of the fossil plant Rhynia major. *This primitive vascular plant grew an underground rhizome that bore upright aerial stems. The plant had no leaves but its stems were dotted with spiny hairs.*

SPORE-BEARING PLANTS GIVE RISE TO THE FIRST SEEDS

The period between 394 and 299 Ma (mid-Devonian to the end of the Carboniferous) saw land plants evolve from diminutive nonvascular and vascular plants into tall trees. A large expansion in the number of species took place at the start of this period, trebling the number of spore-producing plants. At the same time, vascular plants evolved to become larger with more complex systems for transporting water and nutrients.

Fundamental to this was the evolution of three types of stele, the core cylinder that transports water and nutrients around plants. The formation of wood and bark, and the establishment of root systems, also enabled trees to remain stable while growing much taller. The earliest land plants had no leaves; for 40 million years they photosynthesized, instead, from their stems or spinelike attachments. The fossil record shows that this situation came to an end between 390 and 354 Ma (mid- to late Devonian) with the evolution of leaflike structures and true leaves optimized for photosynthesis. Two leaf types appeared: microphylls (having a single, unbranched strand of vascular tissue); and megaphylls (with multiple or branching veins). These differences still exist in plants today.

At the same time as plants gained the capacity to grow into tall trees, they evolved from producing spores of one size in their sporangia to creating spores of two sizes (known as megaspores and microspores). The larger megaspores developed into female gametophytes, which produced egg cells, while the smaller microspores developed into male gametophytes that produced sperm cells used to fertilize the egg cells.

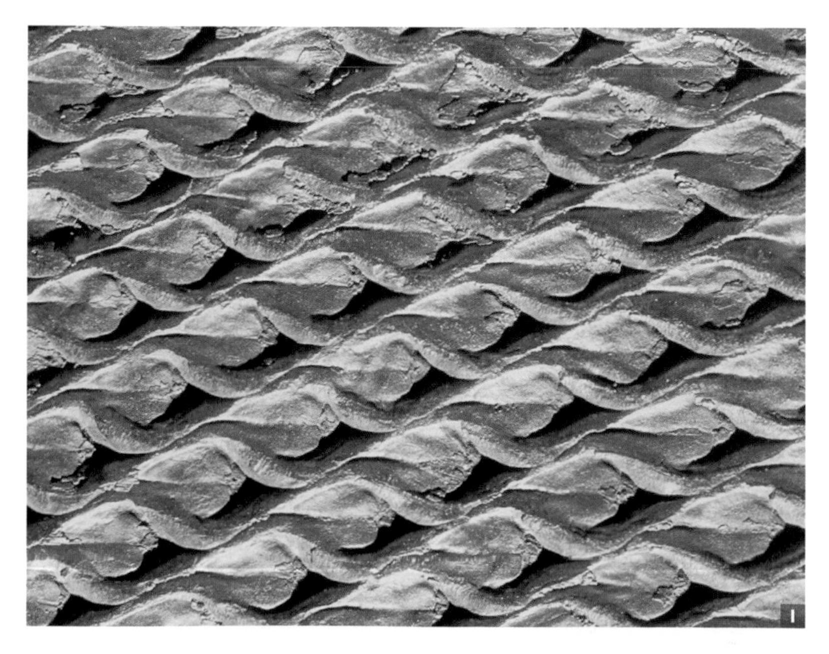

1 *A magnified image of the trunk of a fossil clubmoss tree. The diamond-shaped scars show where leaves once grew from the trunk. Giant clubmoss trees were common 345–270 Ma.*

2 *A fossil of a seed fern (an extinct group of seed-bearing plants) of the genus* Neuropteris, *dating from the Carboniferous period.*

2

EVOLUTION OF THE SEED

When spore-producing plants release their spores, they are at the mercy of the environment when it comes to survival. Over time, the female side of the sporangium went from producing many spores to producing just one. Eventually, this single spore, instead of being dispersed from the sporangium, was retained by it, the sporangium itself becoming enclosed by a protective layer called the integument to form an ovule. "This is, in essence, [once fertilized] the seed," explains Peter Crane. "Keeping the spore attached to the plant gave the opportunity to provide it with nutrients; to provision it. The male spores, meanwhile, still needed to gain access to the egg cells, so they evolved to form pollen."

The evolutionary changes that plants underwent from the mid-Devonian to the end of the Carboniferous corresponded with major environmental shifts. At this time, a supercontinent named Pangaea existed. Global climates were generally warm and humid up until 360Ma, and levels of atmospheric carbon dioxide (CO_2) were high. However, from 360 Ma onwards, the climate cooled and became drier. With land being increasingly vegetated by larger trees, now well equipped for survival with seeds and pollen, early ecosystems continued to develop. By the end of the Devonian, tall trees resembling modern ones existed in large numbers. Then, during the Carboniferous, great swamps supported vast forests of spore-producing plants known as pteridophytes, which included 130 feet (40 m) high lycopods, towering tree ferns, and giant horsetails. Because pteridophytes require sperm cells to swim through water on plant surfaces to reach eggs, they did not fare well when the climate became drier, with many becoming extinct at the end of the Palaeozoic era.

Being able to reproduce in drier conditions proved to be greatly beneficial. By the middle of the Carboniferous, seed-bearing Cordaitales were thriving alongside "seed ferns" or pteridosperms. Now extinct, these woody plants may have given rise to the conifer plant family that is still prevalent today. Soon, the lycopods were in decline and conifers were in the ascendancy; modern conifer groups such as *Podocarpus*,

CYCADS' ANCIENT PEDIGREE

Fossils exhibiting close similarities to living cycads date back to 318 Ma, making them the most ancient lineage of living seed plants. However, all current species of cycads date back to a radiation that occurred 12 Ma. Kew Gardens' oldest pot plant is the cycad *Encephalartos altensteinii*, which arrived there in 1775, two years after it was dug up in South Africa by Kew's first official plant collector, Francis Masson.

Araucaria, cypress, and pine date back to this period. New groups of seed plants also emerged: the cycads, ginkgos, and *Bennettites*. The rise of these large seed-bearing plants would have had a major impact on the cycling of carbon by sucking carbon dioxide out of the atmosphere as they photosynthesized and releasing oxygen back into it. Some scientists believe that during the Carboniferous there was 50 percent more oxygen in the air than there is today and that this could explain the presence in the fossil record of giant dragonflies and oversized amphibians.

PLANT–ANIMAL RELATIONSHIPS

Early animals were primarily decomposers, feeding off detritus left by plants. However, as the Carboniferous advanced, insects began to eat the plants themselves, which would have promoted natural selection of plants with tougher plant tissues and perhaps early chemical defense mechanisms. Today, many living plants produce chemicals to deter pests, and it is possible that these mechanisms began evolving as soon as insects became plant predators, though admittedly it is not possible to detect this in the fossil record. By the Triassic period, vertebrates were eating seeds and dispersing them through their guts. A little later, animal pollinators emerged; there is evidence of pollination by beetles from this time.

Global ecosystems continued to evolve during the Jurassic, when dinosaurs came to predominate, many of them evolving long necks that helped them reach tasty plant material. Selection would have worked strongly to promote the survival of plants that benefited from these emerging interactions, a process that culminated in the evolution of the flowering plants, which took plant–animal relationships to a whole new level.

1 *Cycads, illustrated here growing in Durban Botanic Gardens, South Africa, are the most ancient lineage of living seed plants.*

2 Cycas rumphii, *a species of cycad that grows in Indonesia and Papua New Guinea.*

FLOWERING PLANTS QUICKLY GAIN GROUND

1 *Scientists believe* Amborella trichopoda, *seen here, is the most primitive extant angiosperm; in other words, it forms the lowest branch of the family tree of flowering plants.*

2 *The genus* Nymphaea, *the water lilies, are among the most ancient lineages of plants. This is* Nymphaea nouchali, *growing in Botswana's Okavango Delta.*

The angiosperms, or flowering plants, are the most successful plants on the planet. There are 380,000 living species, ranging from tiny floating duckweeds to the tallest of flowering plants, the 1,230 feet (375 m) high giant ash. Collectively, they represent 96 percent of all terrestrial vegetation, whereas the non-flowering but seed-bearing gymnosperms number fewer than 1,000 living species. The emergence of flowers as reproductive organs around 130 Ma and their subsequent evolution; the shifting of continents and climates through time; and ever-increasing interactions between plants and animals, during the Cretaceous period and beyond, are all likely to have contributed to the angiosperms' great success.

Innovation in sexual reproduction certainly gave flowering plants the edge. When reproduction takes place in a flower, sperm cells travel within the pollen tube through a closed carpel to reach the ovule, whereupon a process called double fertilization takes place. One sperm fertilizes the egg, while another fertilizes a neighboring cell to form a food supply for the embryo. This joint fertilization process, unique to angiosperms, ensures that embryo and food grow contemporaneously. Moreover, the plant's "larder" is only produced once fertilization has occurred, in contrast to gymnosperms, whose seeds are loaded with nutrients prior to fertilization. Consequently, gymnosperm reproduction takes longer and wastes nutrients stored in seeds that are not fertilized. Angiosperms' faster breeding and efficient use of resources would have helped give them a competitive advantage as they evolved.

DNA analysis has helped botanists construct a family tree of angiosperms. Their studies show that *Amborella*, an evergreen shrub with small cream flowers and red berries, is the most primitive extant angiosperm. Water lilies, grasslike *Hydatella*, and a group of shrubs called Austrobaileyales (of which the spice star anise is a member) also have an ancient lineage. Indeed, given that *Hydatella* species nourish their seeds before fertilization in the same way as gymnosperms, it could be that this genus forms the missing link between the non-flowering, seed-bearing gymnosperms and the flowering angiosperms. And though it is possible that the very first flowering plants emerged as far back as 180 Ma, most angiosperm lineages appeared later, in the late Jurassic, in a sudden flurry of evolutionary activity that produced all the major lineages of living flowering plants.

FLOWERING PLANTS DIVERSIFY

Scientists have long been puzzled as to why flowering plants became so diverse, so quickly; Charles Darwin described their rapid expansion as an "abominable mystery." The repositioning of continental landmasses and the resulting changes in climate over the past 130 million years may certainly have contributed. When flowering plants emerged, concentrations of CO_2 in the atmosphere were higher, the continents were much closer together and there were no low-latitude wet rainforest belts. Over time, however, the continents dispersed, CO_2 levels fell, and the climate evolved to include an ever-humid band around the planet's girth, along with cold polar regions. The wet tropics gave rise to the rainforest ecosystems we see today, while the chilling at the poles helped to form the glaciated landscapes of Antarctica and northern latitudes. "Since the heyday of the gymnosperms, the world has become more differentiated geographically and ecologically," explains Peter Crane. "As a result, angiosperms have evolved within a more complicated environment than earlier plants."

POLLINATION AIDS DIVERSIFICATION

Increasingly complex relationships between plants and animals, produced by natural selection, are also likely to have played a role in the development of flowering plants. Fossil pollen suggests that three-quarters of these plants were being pollinated by insects as far back as 96 Ma, a proportion that is similar to today. Plants lure insects to them by offering the food reward of nectar. When an insect inserts its proboscis into a flower to take out the nectar, it inadvertently takes pollen away with it. Then, when it next visits a plant of the same species, it pollinates that plant. It is beneficial for a plant species to forge a relationship with a specific pollinator, because although this means it will have fewer pollinators, the insects that do come will only go on to visit flowers of the same species. As a result, less of the plant's pollen will be wasted, given that most plants are infertile to the pollen of other species. Moreover, the adaptation also benefits the insect, as it is not forced to compete with other insect species for that particular plant's nectar.

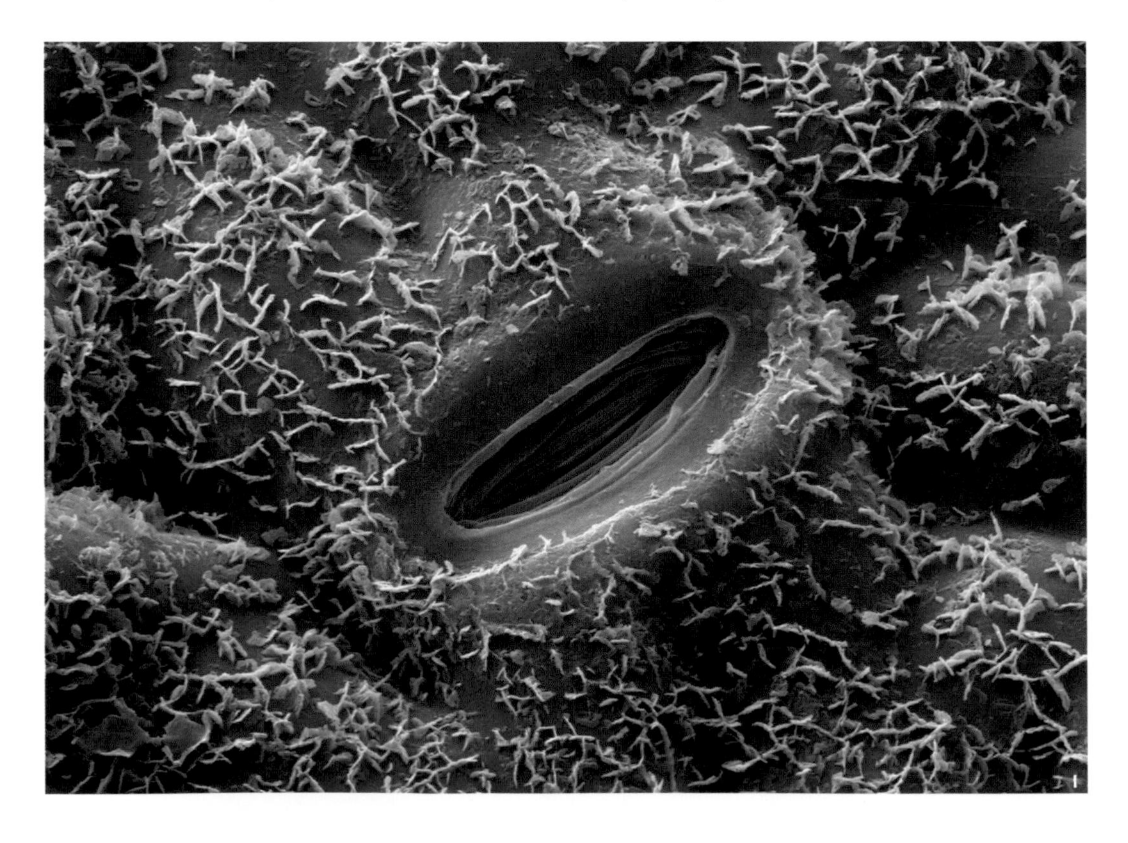

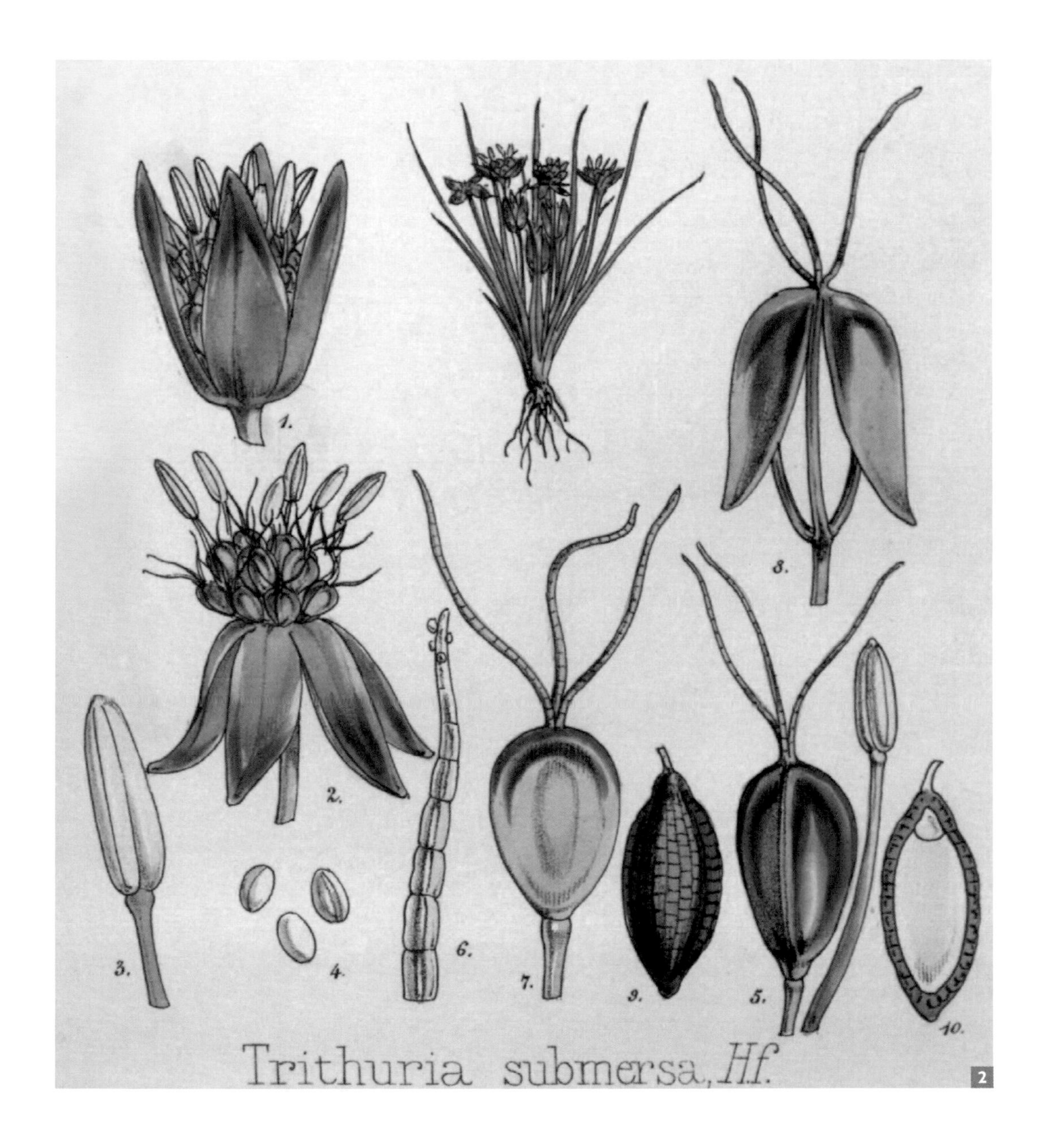

1 *A scanning electron microscope image of an oak leaf stoma. Plants take CO_2, used during photosynthesis to create food, into their leaves through their stomata.*

2 *Scientists believe that plants of the genus* Hydatella *(which was subsumed into the genus* Trithuria *in 2008), could be the missing link between non-flowering seed-bearing gymnosperms and flowering angiosperms, as they nourish their seeds before fertilization in the same way that gymnosperms do.*

Not all angiosperms rely on pollination by animals; the grass family (Poaceae), for example, is the fifth-largest plant family on Earth but is entirely wind-pollinated. So other factors lie behind the angiosperms' success. In the case of grasses, these include being able to tolerate grazing and fire; being able to reproduce asexually as well as through pollination; and effective seed dispersal.

Angiosperm success may also be related to efficient photosynthesis: a report released in 2009 found that later angiosperms had more veins on their leaves than have been found on those of more primitive plants. To grow, plants take in CO_2 through valves in their leaves called stomata. A downside is that these valves lose water when open. The scientists found evidence that angiosperms evolved to have a greater density of leaf veins, while non-angiosperms did not. This would have enabled the flowering plants to absorb more CO_2 while remaining hydrated, at a time when atmospheric levels of CO_2 were falling. The report's authors found that a tripling of vein density would have doubled photosynthesis, providing more carbon for growth.

SURVIVAL OF ANGIOSPERMS

The reason for angiosperms' success, therefore, lies in their ability to exploit different opportunities, be they for creating new species, reproducing more efficiently, dispersing seeds effectively, or outgrowing their competitors. Their diversity and ability to inhabit different environments have meant that extinction among the angiosperms has been low and their ability to create new species has been high. Plants in general seem to be relatively resistant to mass extinctions; the angiosperms emerged relatively unscathed from the mass extinction at the end of the Cretaceous that saw off the dinosaurs and other animals.

However, whether the flowering plants can evolve sufficiently to withstand the challenge of human-induced climate change remains to be seen. Given that plants have evolved over millions of years and the expected shifts in climate are likely to take place over a few centuries, or even a few decades, the plant world may yet be facing its greatest evolutionary test.

MASS EXTINCTIONS AFFECT ANIMALS MORE THAN PLANTS

The Earth has experienced five mass extinctions in its history, where abnormally high numbers of species died out in a short time period.

Ordovician–Silurian (443 Ma)
Around 85 percent of sea life was wiped out when a huge southern ice sheet prompted sea levels to fall and changed the ocean chemistry.

Late Devonian (359 Ma)
Three-quarters of organisms became extinct, primarily those that lived in shallow seas. Changes in sea level, asteroid impacts, climate change, and the emergence of plants may have contributed.

Permian (248 Ma)
This event is called the "Great Dying" as 96 percent of all living things died out. All life on Earth today is descended from the surviving four percent. Marine creatures and insects were most badly affected by the extinction, which has been blamed on asteroid impacts, eruptions, a large release of methane, a drop in oxygen concentrations, sea-level fluctuations, or a combination of these.

Triassic–Jurassic (200 Ma)
Half of all species became extinct including marine reptiles, large amphibians, reef-building creatures, and cephalopod molluscs (a group of sea animals that includes the living *Nautilus*). Plants were not badly affected. Potential reasons for the mass extinction include climate change, eruptions, and an asteroid impact.

Cretaceous–Tertiary (65 Ma)
Known as the K/T extinction, this brought about the demise of the dinosaurs, ammonites, some flowering plants, and the pterosaurs (reptilian close cousins of the dinosaurs).

1 *Star anise* (Illicium verum), *often used to flavor Chinese food, is a member of the Austrobaileyales order, one of three most ancient branches of flowering plants.*

2 *The star-shaped fruit comprises around six to eight single-seed pods, which radiate from a central point.*

THE RISE OF ANNUALS UNDERPINS HUMAN SUCCESS

Scientists have turned to genetics to try to resolve Darwin's "abominable mystery," namely, why angiosperms have been able to diversify so rapidly. When organisms reproduce sexually, in order to produce ova (egg) and sperm cells (in the case of plants, pollen) they undergo a process called meiosis. During this process, the number of chromosomes (structures of DNA that carry genetic information, or genes) in each cell is halved from two (diploid) to one (haploid). When these cells then fuse together during fertilization, they create a new plant in which the original number of chromosomes is restored. Sometimes, however, a failure in meiosis results in a plant that is polyploid; that is, it has more than two sets of chromosomes. Geneticists believe polyploidy may provide the key to the angiosperms' success.

POLYPLOIDY MAKES A DIFFERENCE

If a diploid cell fuses with a haploid cell, the resulting plant will have three chromosome sets (triploid) rather than the usual two; and if it fuses with another diploid cell, the offspring plant will have four sets of chromosomes (tetraploid). This does not happen often, but over the course of long evolutionary timescales it happens often enough to produce a noticeable effect. In some groups of plants this process of aggregating chromosomes continues, so if one of these tetraploid cells fuses with a diploid cell, it can produce a hexaploid and so on. The highest level of polyploidy so far recorded in a flowering plant was 80 sets of chromosomes, encountered in a stonecrop (*Sedum*) from Mexico. "We now know that there was a polyploid event that took place before angiosperms and gymnosperms split, and another that was shared by all angiosperms," says Mark Chase, Keeper of the Jodrell Laboratory at Kew Gardens. "The plant *Arabidopsis thaliana*, meanwhile, [the first plant to have its entire genome sequenced] has had up to six rounds of polyploidy in its history."

CONTROLLING GENES

Studies of *Arabidopsis* threw up a paradox. Although it has only five pairs of chromosomes, it has high numbers of genes that control other genes (those that might dictate whether a particular gene to make flowers pink was expressed, for example). The low number of chromosome pairs suggested that it had not been subjected to many rounds of polyploidy, whereas the high numbers of controlling genes suggested it had undergone six rounds. When the scientists delved into this conundrum further, they found that many plants with low chromosome numbers had high numbers of controlling genes; and when they looked still closer, they realized that these plants, after a polyploid event, were essentially condensing the number of chromosomes by taking genes from one chromosome and adding them on to the arm of another one. "Any traits carried on that arm would be inherited together with much higher frequency [during fertilization] because they're on the same chromosome 'linkage group,'" Mark explains.

Early angiosperms were much like their seed-bearing gymnosperm ancestors. They were woody plants, and their life strategy was, like modern trees, to live for a long time and produce seeds every year. As long as the conditions for seeds to germinate and thrive were good in some years, the species would survive. In a poor year, though, they had to hold back some resources from seed production to ensure the parent plant survived. With the evolution of annuals, however, came a major shift in angiosperm life strategy. These new kinds of plants, which include grasses such as maize and wheat, tomatoes, many brassicas, and sunflowers, put all their resources into making seeds, as the parent plant can only survive for a single growing season. By studying many different annuals, scientists found that they all exhibit lower chromosome numbers (with their large number of linked genes) and high numbers of controlling genes.

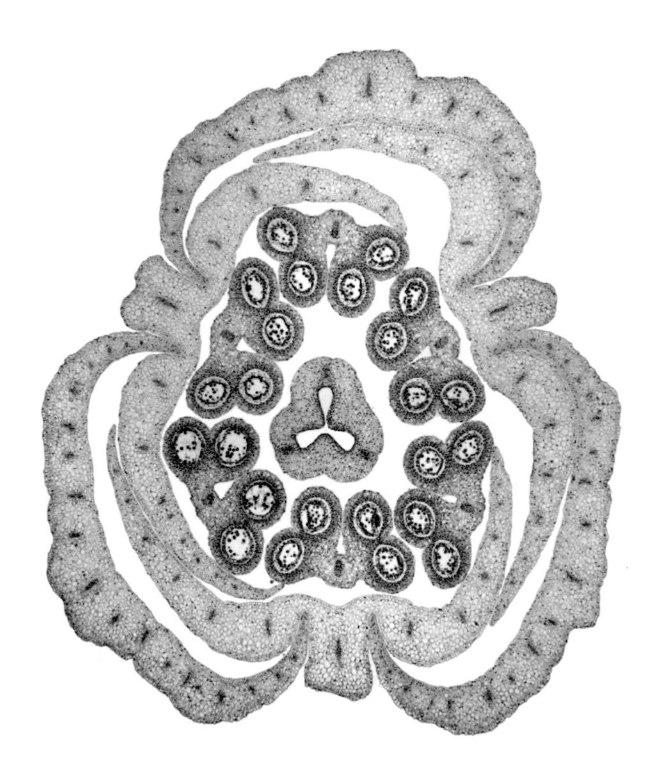

A light micrograph of a lily flower bud, showing pollen cells (circles) dividing within the anthers (darker inner ring). The anthers have four pollen sacs, within which the diploid sporophyte pollen grains are undergoing meiosis to form a haploid gametophyte.

EVOLUTION IN PROGRESS

The plants exhibiting these characteristics also have something else in common: novel evolutionary strategies compared with longer-lived perennials. For example, in the Oncidiinae orchids (a subtribe of around 70 genera), most species take seven years to go from seed to flowering, live on the main axes of trees, where nutrients and water are easy to come by, and have 30 pairs of chromosomes. However, in several groups of orchids within this subtribe, the chromosome numbers are lower (in one case reduced sixfold), they go from seed to flowering in one year, and live on the very tips of the tree branches where water is less readily available. Similarly, Australian tobacco plants that are recent polyploids with higher numbers of chromosomes tend to live on the north and east coast in relatively moist areas. Those with lower chromosome numbers appear to be moving into the deserts, a great number of them (12 of the 26 species native to Australia) close to Uluru.

We still have much to learn about the process of how some angiosperms evolved to become annuals and the ongoing evolutionary path that will bring about novel plants in future. What is interesting, however, is that if angiosperms had not evolved rapid-cycling annuals that put all their energy into seed production, they would not have been useful to humans as crops. It is not possible to get sufficient food resources from perennials, so without annuals civilization might never have got off the ground. The genetic work conducted by Mark and other scientists has revealed that only plants with low chromosome numbers can complete the seed–flower cycle rapidly. "We think that by controlling gene expression in a much more sophisticated way, plants can deal with environmental extremes and still live successfully, which is likely to have contributed to their success," Mark says. "Getting insights into what has made these plants become so dynamic should help us to breed future crops that have the capacity to be more resilient."

HOW PLANTS EVOLVED FROM ALGAE TO ANGIOSPERMS

4,000 MA
Life on Earth emerges in the form of tiny microbes gathered around subsea geothermal vents.

3,500 MA
Photosynthesizing organisms appear.

2,700 MA
By now, organisms exist incorporating mitochondria (energy-producing organelles) and chloroplasts (photosynthesizing organelles).

1,870 MA
The first fossils of multicelled (eukaryotic) organisms date to this time.

450 MA
The first conclusive evidence of land-dwelling plants, in the form of spore-containing sporangia of bryophytes (a group that includes mosses), found in Oman, has been dated to this time.

440 MA
Vascular plants, those with lignified tissues called xylem that circulate water and nutrients around their structures, appear in the fossil record at this time. Classified as belonging to the genus *Cooksonia*, these early plants have small, divergent stems topped with sporangia.

400-412 MA
By now, many small vascular plants live on several continents. The first horsetails appear. The Rhynie chert, containing exceptionally well-preserved terrestrial plants and animals dates to this time. The flora includes seven named higher land plants, all less than 16 inches high: *Rhynia, Aglaophyton, Asteroxylon, Nothia, Horneophyton, Trichopherophyton,* and *Ventarura*. Plants begin to exhibit two types of spores: megaspores and microspores. The oldest fossil insect, *Rhyniognatha hirsti*, dates to around this time.

395 MA
Land plants begin to grow taller, and the number of spore-producing plant species increases. The first lichens and stoneworts appear. Plant vascular systems become more advanced; the evolution of roots, wood, and bark facilitates taller growth.

390 MA
True leaves that can photosynthesize first appear in the fossil record at this time.

385 MA
The first treelike plants (progymnosperms) appear and forests begin to develop. The earliest fossil tree (*Pseudosprochnus*), dating to this time, was found in New York in 2007. Also known as Gilboa, the trees had 26 feet- (8m-) high trunks topped by leafless photosynthetic branches.

383 MA
The progymnosperm *Archaeopteris*, a tall tree with spreading branches, forms the first global forests.

370 MA
The first fossil megaspores appear at this time, with an outer coating that would have protected the megasporangium from drying out and attack. As such, they represent the evolution of seeds. The earliest seed plants are seed ferns, a now extinct group, resembling ferns, which reproduced by means of seeds that grew along their branches. The earliest fossil seed fern is *Elkinsia polymorpha*.

360 MA
Seed ferns become widespread. Together with spore-producing giant horsetails, lycopsids, filicopsids (ferns), and progymnosperms, they form dense swampy forests across the globe. The warm, humid global climate begins to cool and become drier. The concentration of CO_2 in the atmosphere begins to fall.

320 MA
Conifers and glossopterids (tongue ferns) appear along with lianas and epiphytes. The first mammals appear.

310 MA
Pollen is present in the fossil record by now.

300 MA
The supercontinent Pangaea, formed from the merging of the earlier landmasses Gondwana and Laurasia, now straddles the Equator. As a result, the global climate warms and becomes more arid. CO_2 begins to rise from 300 parts per million.

139 MA
The first angiosperm pollen in the fossil record dates to this time.

150 MA
The first birds appear.

160 MA
The evolution of angiosperms is likely to date from this time. Scientists believe the flowering plants emerged in the ancient tropics, between 0° and 30° latitude, colonizing higher latitudes in the subsequent 20–30Ma. *Amborella* is the earliest living angiosperm clade.

195 MA
For the next 50Ma, conifers, gingkos, ferns, *Bennettites*, and cycads dominate. Apart from Caytoniales, the seed ferns decline.

20 MA
By now, grasslands have emerged, their success aided by drier climates, wildfires, and the evolution of hoofed mammals.

50 MA
Global climates start to cool and become more arid.

55 MA
Angiosperms (monocots and eudicots) diversify. Species of the genus *Nothofagus* (southern beeches), spread across the Southern Hemisphere.

65 MA
The climate is very warm; annual tropical temperatures are 41–50°F warmer than at present. Sixty percent of plant species, along with many animals including the dinosaurs, become extinct in the K/T extinction.

0.15 MA
Evolution of *Homo sapiens*.

Changes caused by human activity, including climate change, urbanization, agriculture, and logging subsequently put as many as 100,000 species—one fifth of the world's flora—at risk from extinction.

0.9 MA
Steppe floras expand.

290 MA
Conifers and seed plants begin to diversify.

280 MA
Atmospheric CO_2 peaks at 1,500ppm, the highest level in Earth's history. Between now and 260Ma, an array of new seed plants arises, including cycads, ginkgos, and *Bennettites*.

248 MA
Half of all plant species, primarily those inhabiting woodlands, die out in the Permian "Great Dying" mass extinction. A group of seed ferns called the Caytoniales appears. Their characteristics, gleaned from fossils and extant taxa, suggest they are close relatives to the flowering plants (angiosperms).

225 MA
Rapid global warming takes place. Some areas are dry while others experience monsoons. The first Coccolithospores (ocean-living plankton) appear. Cycads, *Bennettites*, and conifers diversify. The lycophyte *Pleuromeia* spreads across the globe, its success aided by the Great Dying. The first dinosaurs appear. Some scientists have suggested that dinosaur feeding behavior promoted the evolution of flowering plants.

129 MA
Early fossil flowers, which are very small, date to this time. Fruits, flowers, and seeds from more than 100 angiosperm taxa dating to this time have been found in deposits in the Vale de Água, Portugal. Contemporaneous almost-complete specimens of early aquatic angiosperms have been found in a lake deposit in China.

100 MA
The rapid diversification of angiosperms begins. Clear fossil evidence of pollination by insects dates to this time; for example, thrips (Thysanoptera), covered in pollen grains, were found preserved in amber in Spain.

70 MA
The first fossil grasses date to this time. Many modern groups of flowering plants are present by now, including the magnolias. Many key pollinators emerge, including bees, and modern moths and butterflies. Showy petals become more common.

15 MA
Broad-leaved evergreen plants and conifers become more widespread in the northern hemisphere; broad-leaved forests decline.

10 MA
Ice sheets now cover the South Pole and are building up in northern latitudes.

2 MA
Conifers diversify at high latitudes.

1.8 MA
Evolution of *Homo erectus*.

EVOLUTION OF LAND PLANTS CULMINATES WITH THE DRAMATIC RISE OF ANGIOSPERMS

The seed-bearing plants (spermatophytes) are comprised of the gymnosperms (non-flowering) and angiosperms (flowering). The illustrations here highlight the great success of the angiosperms.

The illustration opposite includes the 11 major living land plant groups and several extinct groups (indicated by a dagger). The term phylogenetic refers to the evolution of a genetically related group of organisms, as distinguished from the development of an individual organism. The general topology is based on results from analyses of morphological data, showing the physical form and external structure of plants, which are broadly supported by findings from molecular datasets. A clade is a group of organisms believed to comprise all the evolutionary descendants of a single ancestor. Major clades are indicated by a circle.

A paraphyletic group is one that contains organisms descended from a single evolutionary ancestor but which does not include all the descendent groups. Paraphyletic groups, such as bryophytes, are listed along the top. The relationships among the bryophyte groups have been controversial; this topology reflects the current scientific thinking based on analyses of several large molecular datasets.

The number in brackets after each group of extant land plants refers to the approximate number of species in that group. Note the overwhelming diversity of angiosperms (flowering plants) compared to all other groups; approximately 90 percent of all living terrestrial plant species are angiosperms.

THE ANGIOSPERMS OUTNUMBER EARLIER PLANTS

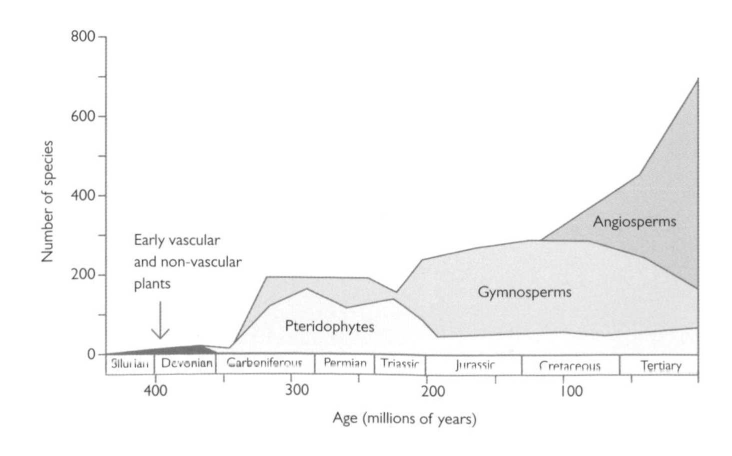

Evidence from the fossil record for the emergence and major expansion of the angiosperms during the Cretaceous and Tertiary.

A PHYLOGENETIC FRAMEWORK FOR THE EVOLUTION OF LAND PLANTS

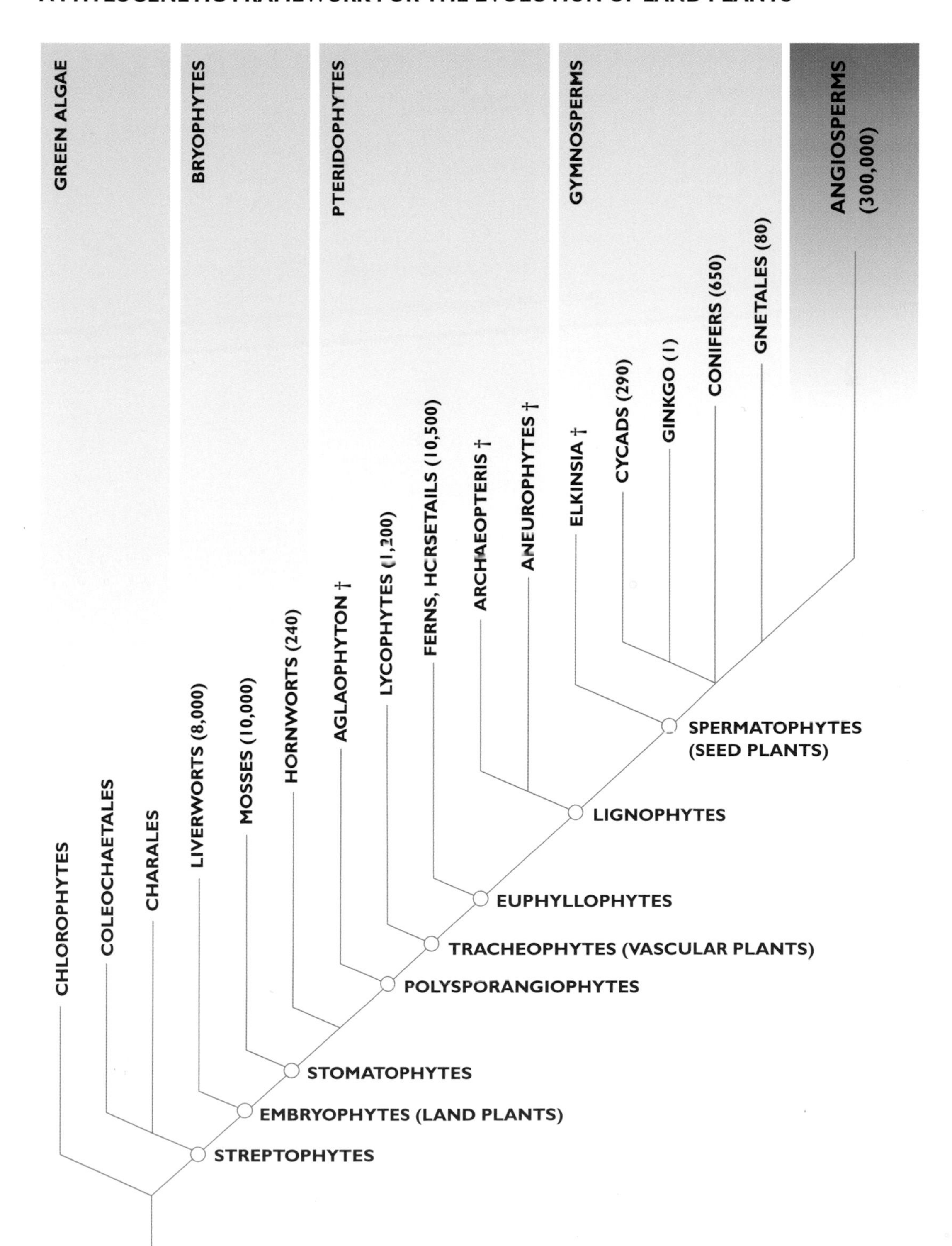

PLANTS AND SEEDS FROM THE WORLD'S RAINFORESTS

The world's tropical rainforests extend in a band around the Equator, bounded by the Tropic of Cancer at 23° north and the Tropic of Capricorn at 23° south. They are particularly diverse areas; 2½ acres (1 hectare) of the Amazon rainforest can contain 1,500 species of higher plants. Many rainforest species are useful to humans but climate change, illegal logging, and urban developments are causing swathes of rainforest the size of Panama to be lost every year. The rainforests of tropical West Africa are some of the most species-rich habitats in the world but are particularly poorly known. Scientists at Kew Gardens are focusing on recording and learning to germinate seeds from threatened West African tropical species so they can pass that knowledge on to locals and help conserve vulnerable species. Here are some important species that face extinction.

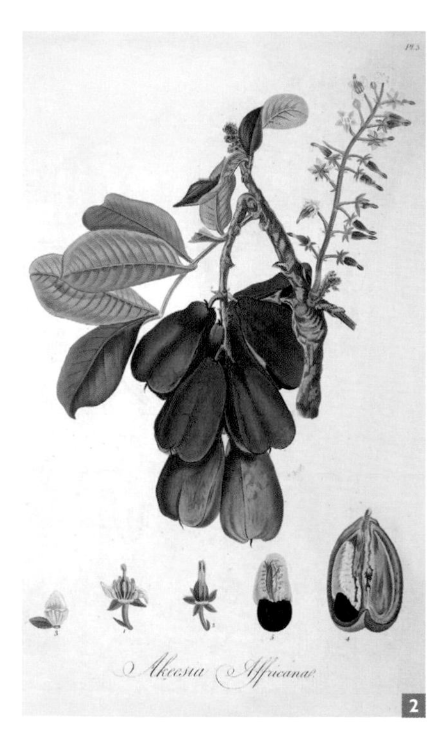

1 *Rainforests, such as the Bossu Forest in Guinea seen here, cover around 6 percent of the world's land area but are home to 50 percent of all living things.*

2 Blighia sapida, *the seeds of which are used to make soap.*

3 Gardenia nitida, *a member of the coffee family, which is used in West Africa to protect dwellings from lightning and to boost women's fertility.*

BLIGHIA SAPIDA

Belonging to the soapberry family and named after William Bligh, captain of the *Bounty*, this fruit tree is native to the Guinean forests of West Africa. Three seeds grow within a fleshy yellow aril, enclosed by a green leathery skin that turns bright orangey-red when the fruit is ripe. Commonly known as ackee, the fruit's aril is edible when mature. The seeds and capsules are poisonous when immature but are used to make soap. All parts of the tree have medicinal properties.

Because of the species' potential to contribute to local livelihoods, it has been earmarked as a high priority for domestication in Benin. However, little is known about its characteristics or how it can best be cultivated. Like many plants from wet tropical habitats, *Blighia sapida's* seeds are sensitive to desiccation. The seeds are probably intermediate or recalcitrant, but some scientists have found that viability can be maintained for three months in moist storage.

GARDENIA NITIDA

This glossy-leaved tree, with fragrant white flowers, is one of the rarest plants in Burkina Faso. Kew Gardens' first Director, William Hooker, first published the name and a description of the species in 1847. However, it wasn't until 2007 that Kew scientists were able to gather seeds from the plant during an expedition organized jointly by Ouagadougou University and the Floristic Center of Abidjan, Ivory Coast. Even then it was not an easy task: four hours of walking in temperatures of 95°F (35°C) yielded only five individual plants growing in dense forest along a 1¼ mile (2 km) stretch of the Comoé River.

The seeds of this plant, which is a member of the coffee family, are thought to be orthodox. In experiments, the seeds survived drying without a significant reduction in their viability; it should therefore be possible to bank them. In germination tests conducted at the MSBP, 75 percent of dried seeds germinated at 77°F (25°C).

Being able to cultivate *Gardenia nitida* would benefit communities across West Africa, where it is sometimes planted for its sacred and protective properties. In Ghana, its wood is put on the roofs of houses to ward off lightning strikes. Meanwhile, women there eat its roots ground up with boiled eggs to boost their fertility.

1 Garcinia kola, *the seeds of which are seen here, is threatened by habitat loss and the felling of trees to make chewing sticks.*

2 *Seeds of* G. kola *inside the fruit.*

GARCINIA KOLA

This evergreen tree, found across West Africa, grows to around 50 feet (more than 15 m) high. Known as bitter kola, it is primarily used to produce "chewing sticks" that help with dental hygiene. However, it is also used for medical purposes, including treatment for worms, tumors, and skin infections.

The MSBP has ascertained that the seeds are recalcitrant, which means they cannot, in theory, be banked. They also have some dormancy; overcoming this requires scientists to mimic the natural environmental conditions in which *Garcinia kola* grows. Joseph Asomaning from Ghana, under the supervision of the MSBP, worked for his PhD thesis on improving germination methods for *Garcinia kola*. Its favored habitat is dense rainforest, beside rivers and in swamps, on granitic soil at altitudes of between 650 and 4,100 feet (200 and 1,250 m). It is listed as vulnerable on the International Union for Conservation of Nature's (IUCN's) Red List of Threatened Species, its existence endangered by general habitat loss and felling to make chewing sticks.

WIDDRINGTONIA WHYTEI

This critically endangered species, also known as the Mulanje cedar, grows only on Mount Mulanje in Malawi. Its extent is believed to be around 2,000 acres (845 hectares). Between the 1890s and 1955, the cedar was exploited commercially; its fragrant, termite-resistant timber was used to make door and window frames, along with craft items. Today, felling *Widdringtonia whytei* trees is illegal, but continued logging, fires, a lack of regeneration, the impact of invasive species, exotic pests, and overcollection of firewood are forecast to reduce its population by more than 80 percent by 2030.

Very little is known about how the plant regenerates naturally, making conservation difficult. Tembo F. Chanyenga of the University of Stellenbosch in South Africa studied the seeds as part of his PhD thesis, which the MSBP co-supervised. He found that

W. whytei cones depend on moderate to severe fires to help widely disperse the seeds in large quantities. However, fires burning outside of forest patches during the dry, hot season also kill seedlings that have germinated in open ground. The plant is a pioneer species needing light to become established, so its seedlings cannot thrive where there are thick litter layers and shady conditions. Saving the tree from extinction may therefore require moderate to severe fires to encourage seed dispersal and exposed mineral soil to ensure seedlings become established.

ALLANBLACKIA SPP.

Allanblackia is a genus of nine species that grow in the tropical forests of East and West Africa. The seeds of these species yield good-quality oil, which experts believe has the potential to provide a healthier alternative to palm oil. Extracted from the fruit of the palm *Elaeis guineensis*, palm oil is used in everything from soap to biscuits, but rapid rise in demand is causing serious impacts in the tropical countries that have established oil palm plantations. In an effort to investigate *Allanblackia*'s potential, Unilever has set up local teams in Ghana, Nigeria, and Tanzania to establish a supply chain for producing oil from wild-harvested seeds. However, extracting the oil in greater quantities would call for cultivation of the plants on a much larger scale.

Initial research showed that seeds of *Allanblackia* species take over ten months to germinate, making cultivating the plants a slow prospect. For example, seeds of *A. parviflora* only germinated seven to 24 months after sowing, while the proportion of seeds that germinated ranged between zero and 35 percent. However, when Moctar Sacande, International Programme Coordinator for Africa at the MSBP, was given the task of trying to germinate *A. floribunda, A. parviflora,* and *A. stulhmannii* from, respectively, Cameroon, Ghana, and Tanzania, he managed to obtain germination in 40 percent of plants within a month. His method involved sectioning and scarifying the seeds and also applying the plant hormone gibberellic acid, which promotes growth.

"I managed to crack the germination issue; I found it is possible to speed up the germination," Moctar says. "The method is relatively low-tech, so what we have achieved should be transferable to farmers in Africa."

Allanblackia *species, such as* Allanblackia floribunda *seen here, yield high-quality oil that could provide an alternative to widely used palm oil.*

S E E D B A N K A U S T R A L I A

AUSTRALIA'S PLANTBANK HELPS TO GROW "DIFFICULT" RAINFOREST SEEDS

1

NAME

The Australian PlantBank, at the Australian Botanic Garden, Mount Annan, Royal Botanic Gardens and Domain Trust, Sydney.

NUMBER OF ACCESSIONS

10,427

WHEN FOUNDED

1988

FOCUS OF THE COLLECTION

The Australian PlantBank holds seeds from around 50 percent of threatened species and 45 percent of known plant species in New South Wales.

WHY IT IS NEEDED

Of Australia's 25,000 species, 23 percent are under threat. In New South Wales there are 5,810 species, 614 of which are listed as threatened species.

WHO FUNDS IT

The Australian PlantBank is one of many cogs in the great global seed banking effort set in motion by the MSBP. It has received funding from the MSBP, the state government of New South Wales, and HSBC Bank Australia.

1 *The Australian PlantBank houses a seed bank, along with tissue-culture and cryogenic-storage facilities.*

2 *Some seeds from each wild collection are tested before banking to check they are able to germinate.*

3 *Tissue culture is used extensively for desiccation-sensitive rainforest species, seeds from which cannot be readily stored in the seed vault.*

WHERE SEEDS ARE STORED

The seeds reside in a state-of-the-art vault within a new building designed by Australian architecture practice BVN and opened in late 2013. It is open to the public, to view and to participate in science and conservation activities. The building has suitably green credentials for a project that seeks to save biodiversity. Designed to capitalize on natural light, warmth, and prevailing winds for passive heating and cooling, it is constructed from insulating, recyclable, and sustainable materials, with water for the toilets and lichen garden saved via a rainwater-harvesting system. The Australian PlantBank integrates the seed-bank, tissue-culture, and cryogenic-storage facilities with Australia's largest collection of indigenous flora. The surrounding Australian Botanic Garden, Mount Annan, displays more than 2,500 plant species across 1,030 acres (416 hectares) of hills and lakes.

CURRENT RESEARCH

Research at the Australian PlantBank focuses on species that are recalcitrant and have proved difficult to germinate and bank. Many eastern Australian species occur in rainforests or habitats with relatively high moisture levels. A large proportion of these species are not orthodox in their seed-storage behavior. The bank's largest project aims to understand the desiccation tolerance of rainforest species and to establish alternative conservation techniques. The Australian PlantBank's scientists also aim to understand the longevity of desiccation-tolerant rainforest species, as many appear to be relatively short-lived in storage. Also of concern are more than 40 species of threatened terrestrial orchids in New South Wales.

SEEDS WITH A STORY

Among the seeds stored at the Australian PlantBank are those from the Nielsen Park She-oak (*Allocasuarina portuensis*), one of Australia's most endangered plants. The distribution of this slender shrub, which bears elongate cones, is limited to Nielsen Park, a small area of Sydney Harbour National Park. There were no plants left at the site where it was originally discovered, but it has now been successfully reintroduced to Nielsen Park and other nearby areas after conservation work by the Australian PlantBank. However, it remains threatened by wildfires, invasion by weeds, and habitat loss.

1 *Visitors can observe the seed specialists at work.*

2 *Checking seed characteristics and germination in the PlantBank physiology laboratory.*

3 *Tissue culture flasks are prepared for use.*

4 *All PlantBank seed collections are fully documented and have matching duplicate herbarium specimens.*

WOLLEMI PINE

(WOLLEMIA NOBILIS)

GENUS Wollemia

FAMILY Araucariaceae

SEED SIZE ¼ inch (4–6 mm), including wing

TYPE OF DISPERSAL Wind. Seed has wings or wing-like features.

SEED STORAGE TYPE No storage behavior is known for genus or species.

COMPOSITION Data not available.

In 1994, David Noble, a park ranger at Australia's Wollemi National Park, spied a tree he did not recognize while abseiling down a sandstone canyon. Collecting seeds was only possible by helicopter, as the cones grew 100 feet (30 m) off the ground, but once that had been achieved, a senior botanist at the Royal Botanic Gardens in Sydney was able to confirm that the tree was an entirely new species of a kind thought to have become extinct millions of years ago. Effectively a "living fossil," it was recognized as the only remaining member of a genus of coniferous trees dating back to the time of the dinosaurs, its closest living relatives being the kauri and monkey puzzle trees of the Araucariaceae family.

The oldest-known of the *Wollemia nobilis*, as the species was named, is a tree called "King Billy" which may be over 1,000 years old. But only around 75 trees remain in the wild, a tiny number that highlights the fragile long-term prospects of this evolutionary survivor. Three years after the initial discovery, the Australian government presented Kew Gardens with two plants and 30 seeds; the plants were grown in the Temperate House and the seeds stored for posterity in the MSBP. Then in 2005, another batch of 30 plants arrived at Kew for hardiness tests. The best growing conditions were found to be shallow, acidic, moist but free-draining soil with low nutrient levels. Seeds were subsequently collected and germinated from cones set in 2011, over 50 of which are now growing in the gardens at Kew and Wakehurst Place, alongside 30 of the original research trees.

Plants of the Wollemi pine, which can grow up to 130 feet (40 m) high, were released for sale in 2005. Not only will these plants help to ensure the species' survival, but the money raised will go towards ongoing conservation research. This is vital, as one of the surviving wild stands has already been affected by the root pathogen *Phytophthora cinnamomi*. The Wollemi pine is now listed under the Commonwealth Environment Protection and Biodiversity Conservation Act 1999, as well as the New South Wales Threatened Species Conservation Act 1995.

1 *The male cone of* Wollemi nobilis.

2 *The Wollemi pine is bisexual, with both male and female cones produced on the same plant. The female cones can be seen here at the top of the tree, with one male cone a little lower.*

CHAPTER 3

HOW SEED PLANTS REPRODUCE

DOUBLE FERTILIZATION BRINGS FLOWERING PLANTS GREAT SUCCESS

1

How seed plants reproduce is a wonder of the natural world, the product of millions of years of natural selection and evolution. Seeds first emerged in the early Devonian, after land plants producing a single type of spore (homosporous) had evolved to produce two types of spore (heterosporous). Homosporous plants, such as mosses and most modern ferns and horsetails, show little sexual differentiation. In heterosporous plants, the smaller of the two spores germinates to produce male gametophytes that produce sperm, while the larger spores germinate to produce female gametophytes that produce eggs. Both homosporous plants and heterosporous non-seed plants require water for fertilization, as the motile sperm swim through water to meet the egg, a hindrance in drier climes.

1 *A moss sporophyte capsule. Mosses are non-vascular and do not form seeds. Instead, they produce thin stalks on which are spore-containing capsules.*

2 *Homosporous sori (clusters of spore-producing sporangia) on the underside of fronds of the common polypody fern (*Polypodium vulgare*). The yellow sporangia are at an early stage of development.*

3 *Moss sporophyte capsules. Bryophytes, which include mosses, liverworts and hornworts, are homosporous; that is, they produce a single type of spore.*

Early heterosporous plants dispersed both small and large spores, but, over time, some plants began to disperse only male spores, retaining female spores on the plant. The evolution of this female spore to a primitive seed took place around 360 to 380 million years ago. In heterosporous seed plants, the female gametophyte is retained on the parent plant (sporophyte) inside an ovule and, in most, a pollen tube enables sperm to travel to the egg inside the ovule without requiring water. All seed plants living on Earth today are heterosporous, having evolved from one of several heterosporous lineages that emerged more than 400 million years ago.

All living seed plants have the same life cycle, in which generations alternate. The first generation comprises a diploid (containing two complete sets of chromosomes, one from each parent) sporophyte. During the process of cell division called meiosis, germ cells of the sporophyte divide into a second generation comprising four offspring cells, each possessing half the number of chromosomes of the parent cell. These haploid spores give rise, by a process of cell division called mitosis, to multicellular haploid gametophytes in which gametes develop. The male gamete is found within the plant's pollen and the female gamete within the ovule. During fertilization, sperm from pollen fuse with an egg cell within the ovule to form a zygote containing a mixture of maternal and paternal DNA. The zygote develops into a multicellular embryo that is enclosed within a protective coating: the seed coat. Through this process of sexual reproduction, and unless polyploidy takes place, plants are able to maintain the same number of chromosomes over successive generations.

Living seed plants are divided into the more primitive non-flowering gymnosperms and the more complex flowering angiosperms. In gymnosperms, male pollen is dispersed, primarily by the wind, from male cones to female cones. The ovule is "naked;" that is, it is not surrounded by any maternal tissue other than the protective cover of the seed coat. The pollen germinates at the entrance to the naked ovule and, in higher (more complex) gymnosperms such as conifers, produces a short pollen tube that delivers the non-motile sperm to the egg, to fuse with it to form the next generation of sporophyte

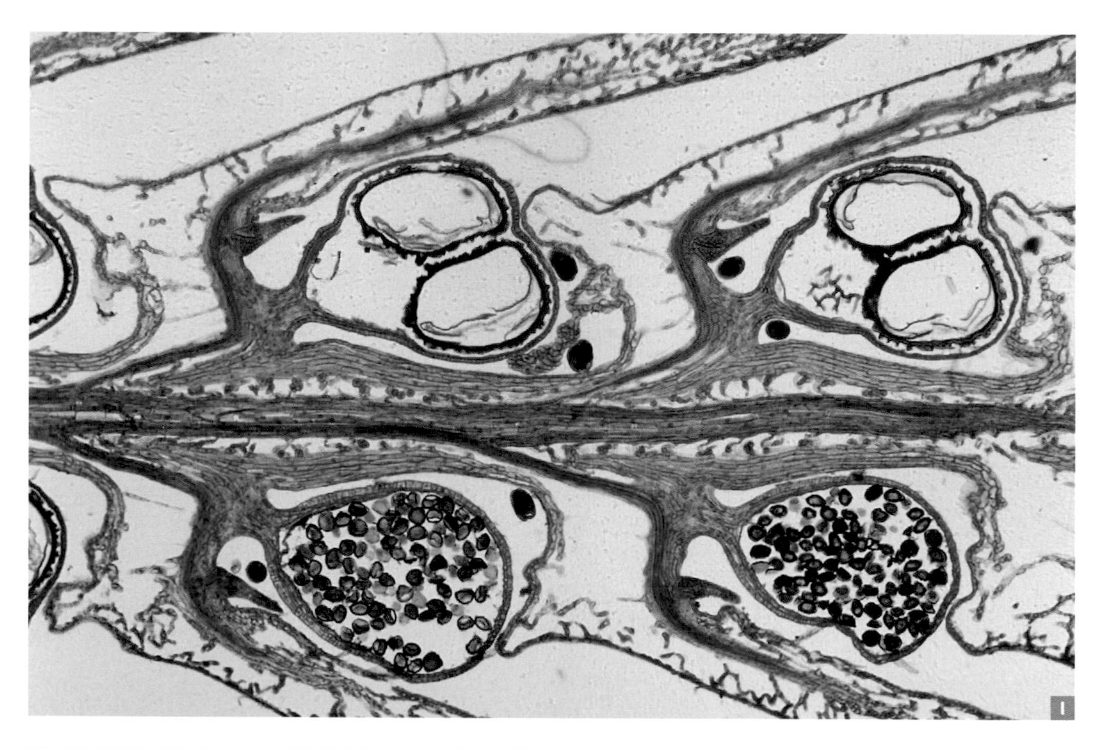

1 *Heterosporous plants, such as this lesser clubmoss (*Selaginella*), produce discrete male and female spores.*

2 *Oregon spikemoss (*Selaginella oregano*) hangs over the trunk of a fallen red alder (*Alnus rubra*) in Hoh temperate rainforest, Olympic National Park, Washington, USA. Lycopods, which include* Selaginella, *evolved some 410 million years ago.*

Pollen being released from Scots pine flowers. Reproduction by gymnosperms, such as conifers, is less efficient than that of angiosperms. Partly as a result, there are fewer than 2,000 species of gymnosperms.

plants. Because gymnosperm plants lay down food resources for the future plant in the ovule before fertilization takes place, if the egg is not fertilized, these food resources are wasted. The fact that gymnosperms constitute barely half of one percent of all extant plant species can be partly attributed to the inefficiency of gymnosperm reproduction.

In contrast to gymnosperms, angiosperms have evolved the less wasteful reproductive process of double fertilization. Pollination takes places when pollen produced by one flower's stamen lands on the sticky platform (stigma) of another flower's carpel. Once the pollen has landed, it grows a pollen tube. The ovule is hidden away within the tissue of the carpel, so a longer tube is needed than with gymnosperms. The pollen navigates its way through this tissue to reach the ovule, inside the ovary, at the swollen base of the carpel. "This is really one of the wonders of flowering plant reproduction," says Simon Hiscock, Professor of Botany and Director of the University of Bristol Botanic Garden. "How the pollen finds its way and navigates through all this tissue—a veritable assault course—in order to deliver the non-motile sperm to the ovule and thence the egg."

On reaching the ovary, one sperm fertilizes the egg within the embryo sac of the ovule to form the zygote that, once fertilized, becomes a multicellular embryo. A second sperm, meanwhile, fertilizes the nucleus of the central cell of the embryo sac. This central cell comprises two haploid "polar" nuclei. The process of fertilization forms a triploid (having three complete sets of chromosomes) fusion product, which develops into the endosperm, the tissue that will feed the zygote as it develops. The fertilized ovule alters to form the woody seed coating that encases the embryo and endosperm. With the process of double fertilization complete, the seed is a plant-in-waiting, with its own larder, and it can remain dormant in its tiny package until favorable conditions for growth come along.

Highly successful, angiosperms constitute 90 percent of all land plants on Earth today. Their unique double fertilization method of seed production, which has endured for millions of years, has been key to their success. "As E. J. H. Corner eloquently explained in his 1964 book, *The Life of Plants*, even an orchid carries a recollection of its marine ancestry," explains Simon Hiscock. "Seed plants still essentially have the same life cycle, albeit highly adapted and derived, as a moss or a liverwort or a hornwort, the most ancient of land plants. Modern seed plants have become better adapted to life out of water but essentially their life cycle evolved in aquatic algae and has been modified through all these millions and millions of years of evolution on land."

1

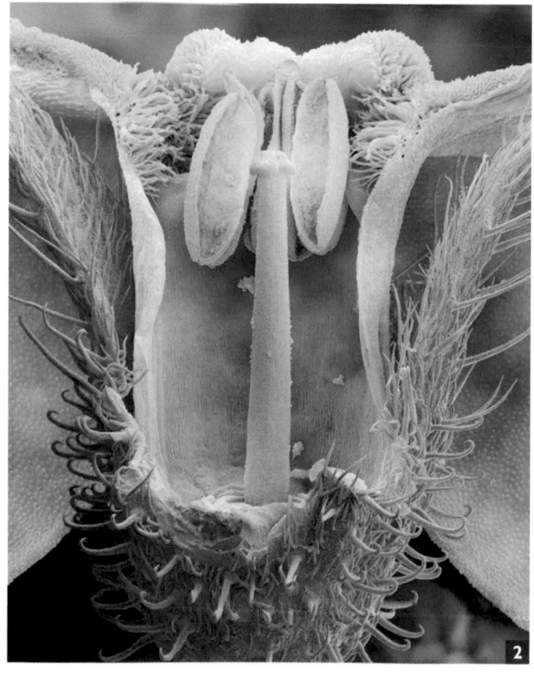
2

3

THE PROS AND CONS OF SEXUAL AND ASEXUAL REPRODUCTION

The process by which seed plants reproduce via double fertilization is sexual reproduction. The offspring inherit genes from both parents, with different offspring inheriting a different set of genes, so that the plant population remains genetically diverse over time. The benefit of diverse populations lies in their greater resilience to pests, diseases, or changing environmental conditions. However, plants that reproduce sexually require pollination for fertilization, and therefore reproduction, to take place.

Some plants are self-pollinating: a self-fertile plant is able to bring pollen and stigmas on a single flower together, avoiding the need for an external pollination mechanism. This can be beneficial in areas where there is little wind or where animal pollinators have disappeared.

Other plants are self-fertile but require assistance to bring pollen from anther to stigma on the same flower. Self-fertile plants pollinated in these ways produce genetically very similar offspring rather than genetically diverse offspring. Therefore, this method of reproduction is only advantageous in very stable environments. Most plants have adaptations that prevent self-fertilization.

1 *This colored scanning electron micrograph (SEM) depicts the pistil (part of the female carpel) of a daisy flower, with pollen (round) produced by the male stamen.*

2 *A colored SEM of a section through a forget-me-not (*Myosotis arvensis*) flower. The sepals (green and hairy) and petals (blue) surround the male and female reproductive structures of the flower. The stamens (yellow) produce male gametes in the form of pollen grains, which are released from the anthers at each stamen's tip. During pollination, pollen grains land on the female stigma (green, upper center).*

3 *This colored SEM shows two pollen grains (yellow) on the stigma (pink) of a goosegrass (*Galium aparine*) flower. A pollen tube is growing from one of the grains. This will extend down through the stigma towards the ovary. Male nuclei travel down the pollen tube to fertilize the ovules and form a seed.*

Many plants are also capable of reproducing asexually. The term "apomixis" is used to describe seed plants that reproduce in this way. Apomictic plants produce seeds from the maternal tissue of the ovule often without the processes of meiosis and fertilization taking place. It is essentially seed production without sex.

A relatively large number of plants appear to be apomictic; the trait has been observed in more than 400 flowering plant taxa, primarily perennials, but it appears to be absent from gymnosperms. Apomictic plants are most commonly found in habitats that are routinely disturbed or where the growing season is short, such as in the Arctic and Antarctic, or where there are other barriers that inhibit the successful crossing of compatible plants.

Another way in which plants reproduce asexually is through vegetative reproduction—for example, through tubers, runners, bulbs, and corms. Apomictic plants often also reproduce vegetatively, usually by growing rhizomes (horizontal underground stems).

When plants reproduce asexually, whether by vegetative means or apomixis, the offspring are genetically identical clones of the parent plant. This makes them more vulnerable to pests, diseases, and changing environmental conditions, although they do have the advantage of not requiring pollination for successful reproduction.

There has been much research in recent decades into breeding the apomictic trait into crops. Currently, farmers spend around US $36 billion each year buying seeds to grow crops. They seek seeds of cultivars that have been bred to exhibit traits that are helpful to them, such as resistance to pests and drought tolerance. The farmers cannot easily grow seeds for themselves because the act of sexual reproduction often eradicates those genes that endow the plant with the selected traits. Only labor-intensive cross-hybridization methods will currently produce robust seeds with the required traits.

In 2010, however, scientists moved a step closer to breeding crops that reproduce asexually using seeds. Researcher Jean-Philippe Vielle-Calzada of the Howard Hughes Medical Institute in the USA discovered that the gene *Argonaute 9* in *Arabidopsis thaliana*, which only reproduces sexually, appeared to "switch off" the initiation of apomixis. The hope is that further studies will one day enable farmers to grow crops with asexual seeds from plants that have already had favorable growing traits bred into them.

HOW GYMNOSPERMS AND ANGIOSPERMS REPRODUCE

The evolution of seed plants in the Late Devonian gave rise to gymnosperms (which include conifers, cycads and ginkgo) and angiosperms (flowering plants). The two differ in their method of reproduction. Gymnosperms use a single process of fertilization in which nutrient resources are stored in the ovule before fertilization. Angiosperms, in contrast, utilize the process of double fertilization, whereby one sperm fertilizes the egg, giving rise to an embryonic plant, while another fertilizes the central cell, to create the endosperm food store.

GYMNOSPERM REPRODUCTION WASTES FOOD RESOURCES

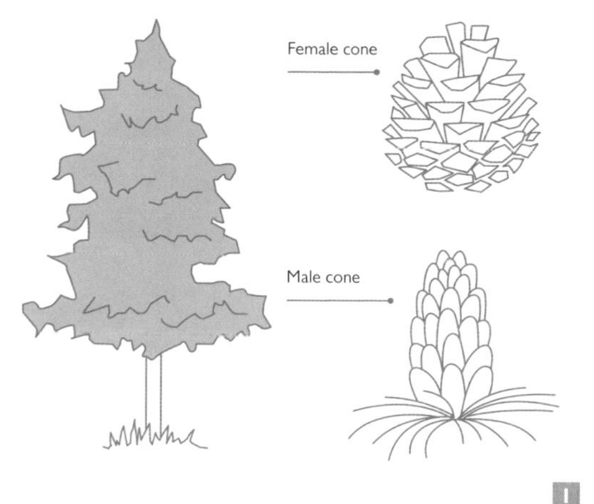

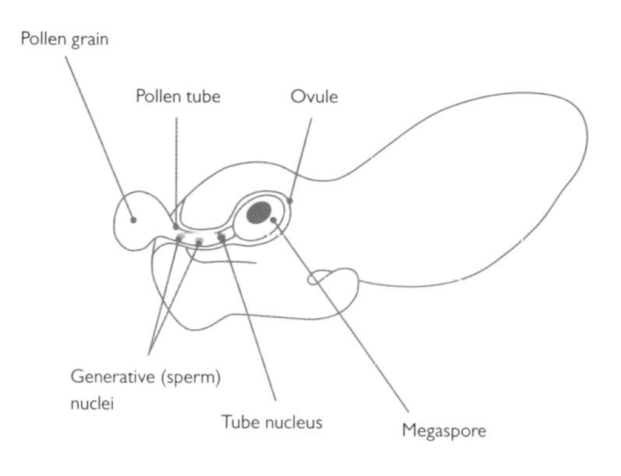

1 *In conifers, male cones grow in the lower branches of the mature tree, while female cones grow in the upper branches. Pollen from male cones is carried on the wind to female cones, where it grows a short pollen tube.*

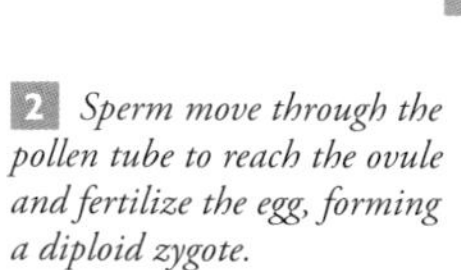

2 *Sperm move through the pollen tube to reach the ovule and fertilize the egg, forming a diploid zygote.*

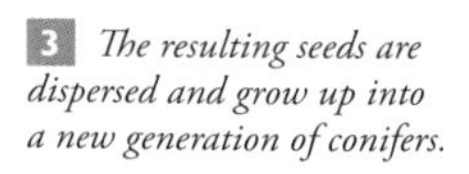

3 *The resulting seeds are dispersed and grow up into a new generation of conifers.*

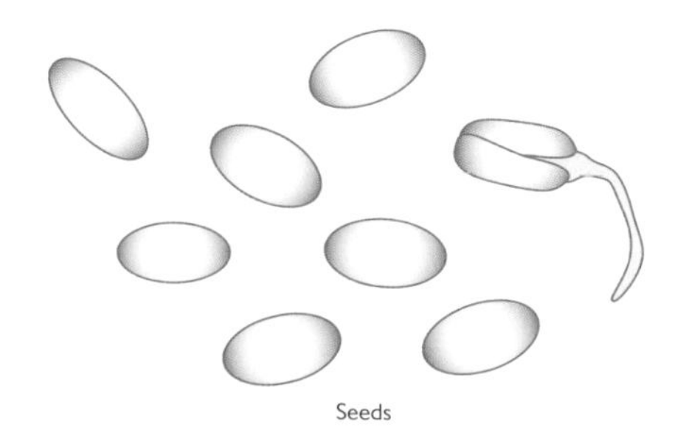

ANGIOSPERMS REPRODUCE EFFICIENTLY VIA DOUBLE FERTILIZATION

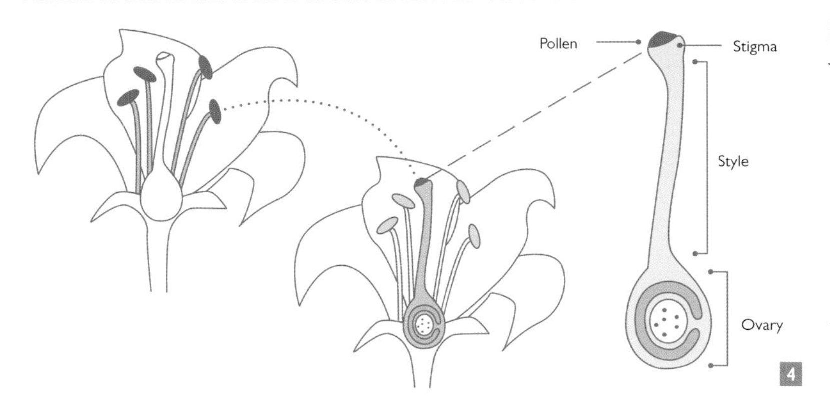

4 *In angiosperms, a plant's flowers bear its reproductive unit, namely stamens and carpels. Pollen is contained within a flower's anthers (male), while the ovule is contained within the ovary at the base of a flower's carpel (female).*

Sometimes there are multiple carpels, in which case this structure is called a pistil. For pollination to take place, pollen must travel from the anther to the stigma on top of the carpel.

5 *Prior to fertilization, the ovule changes. Initially, it contains a diploid cell called the megaspore mother cell. This undergoes meiosis to produce four haploid megaspores.*

In most species, three of these break down to leave one remaining megaspore. This grows and undergoes three phases of mitosis, whereby chromosomes in the cell nucleus split into two identical sets of chromosomes, each with its own nucleus. This results in eight haploid nuclei. The whole structure is called the embryo sac.

The antipodal cells form opposite the opening of the ovule (micropyle). Another three cells form just above the micropyle: Two are synergids, which help to direct the pollen tube, and the other is the egg. Two nuclei remain together at the center of the embryo sac. These are the polar nuclei.

Each pollen grain contains two main cells: the tube cell and the generative (sperm) cell. Once on the stigma, the pollen germinates. The tube cell forms a long structure that moves down the style and into the ovary. The generative nucleus travels behind it, undergoing mitosis to produce two sperm nuclei on the way.

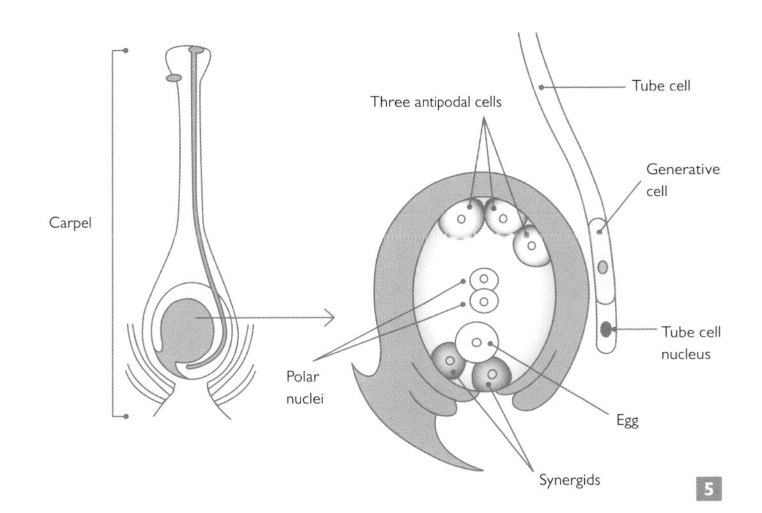

6 *When the pollen tube reaches the micropyle, it sheds the sperm into the embryo sac. One of the sperm cells fertilizes the egg cell to create the diploid zygote that will become the embryonic plant. The other fuses with the two polar nuclei forming a triploid cell that eventually becomes the endosperm. This will provide food for the plant until it is able to produce its own food via photosynthesis. The embryo and endosperm await dispersal, and, once any dormancy has been overcome, they germinate to form a new plant.*

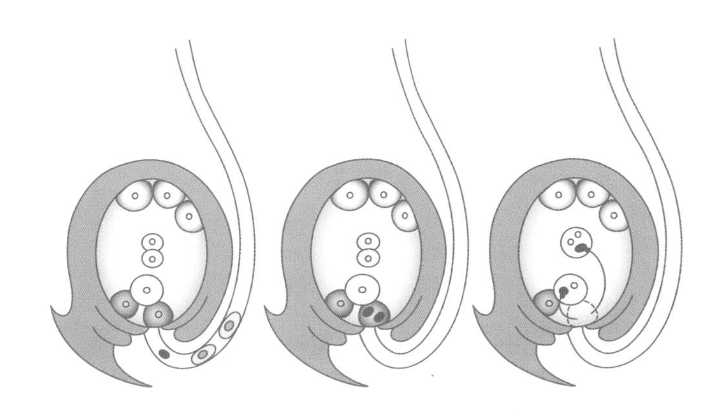

POLLINATION IS A MUST FOR SUCCESSFUL FERTILIZATION

For fertilization to take place, pollination must occur, in which the (male) pollen moves from a flower's anther to the (female) stigma of a carpel. One of the key reasons behind the angiosperms' success is the adaptations they have made, through natural selection, to aid successful pollination. These adaptations include structures that facilitate pollination via the wind, water, or animals. The adaptations that enable animals including birds, bees, moths, possums, and bats to pollinate flowers, often in a way that is mutually beneficial, reflect some of the most remarkable relationships that exist in the natural world.

POLLINATION BY WATER

Only a very small percentage of plants are pollinated by water (hydrophily). Even among aquatic plants, there are many that are pollinated by the wind or by animals. Those that do use water to aid pollination exhibit an array of different adaptations. However, most are pollinated either at, or above, the water surface; few plants are pollinated under water. One of the best examples of pollination by water is that of the aquatic ribbon weed (*Vallisneria spiralis*). In its native Queensland, Australia, it grows in the fast-flowing waters of streams, as well as lakes, ponds, and irrigation channels.

Male and female flowers of *V. spiralis* grow on separate plants. The tiny female flowers float up to the surface on coiled pedicels (small stalks), which also draw them back down into the water after pollination. The pinhead-sized male flowers are concealed in arrow-shaped spathes within the leaf sheath at the base of the plant. Once mature, the spathes open, releasing the male flowers, so that they, too, float to the surface. "Ribbon weed produces these wonderful little floating pollen boats that drift along the surface of the water and then hit the large feathery stigmatic surfaces of female flowers, resulting in pollination," explains Simon Hiscock.

POLLINATION BY WIND

Pollination by the wind (anemophily) is more widespread than aquatic pollination. It tends to be used by temperate trees, such as pines, sycamores, and birches, along with a large proportion of grasses (Poaceae), sedges (Cyperaceae) and rushes (Juncaceae). Many seashore plants and weeds of the goosefoot (Amaranthaceae) as well as the dock

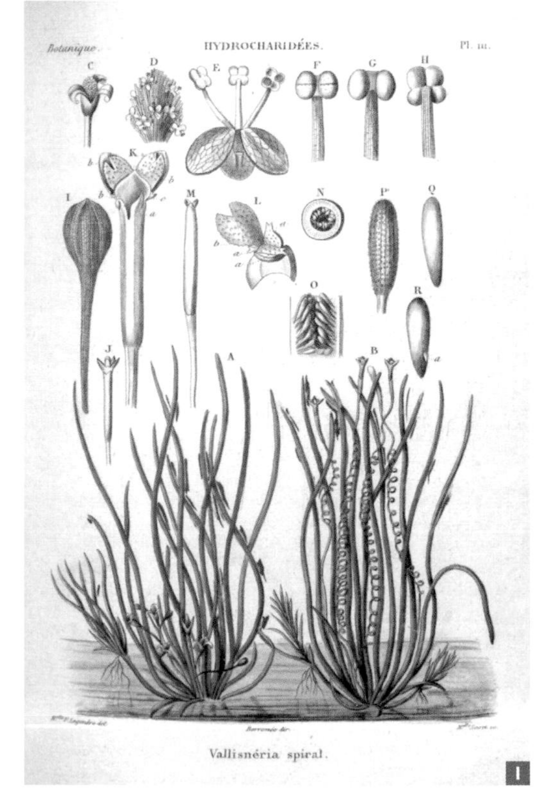

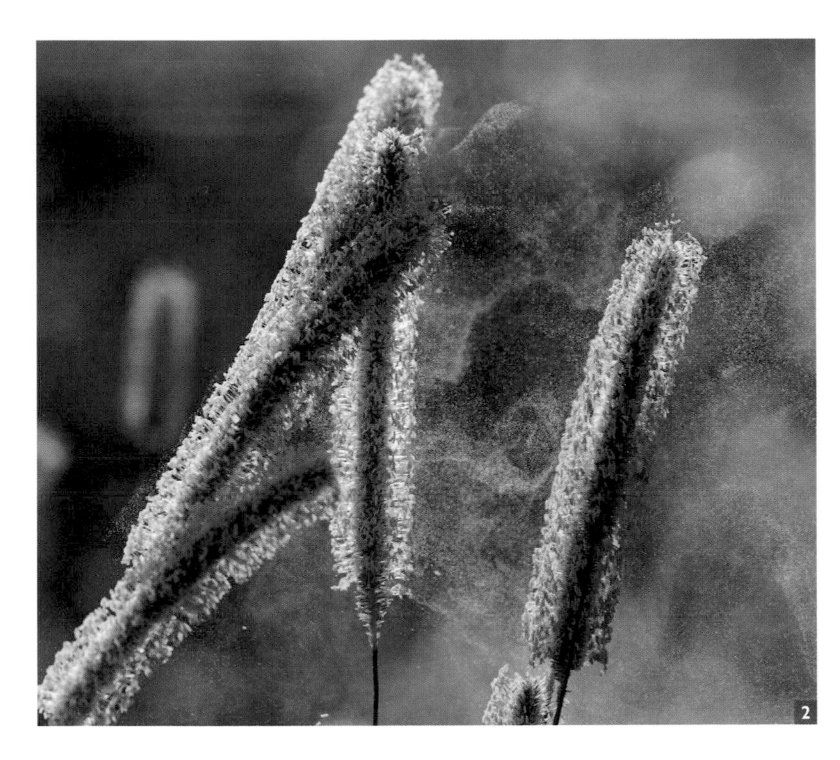

1 *Aquatic ribbon weed (*Vallisneria spiralis*) is one of only a few plants that are pollinated by water.*

2 *Pollen being released from a flowering grass in Vosges, France.*

and sorrel (Rumex) families also engage the wind to disperse their pollen. Although pollination by animals is considered to be the most efficient form of pollination overall, in habitats where wind pollination dominates it is likely that animal pollination would be impractical. For example, insect pollination of abundant grasses on a savanna is unlikely because the ecosystem could probably not support a sufficiently high population of insects during the limited time that the plants are in flower.

Wind-pollinated plants tend to exhibit certain characteristics. For example: their anthers dangle outside of the flowers so pollen can be released more easily; the pollen is very light and smooth to help it drift on air currents; and the stigmas protrude outside of the flowers to trap passing pollen. Wind-pollinated plants usually produce more pollen than insect-pollinated ones, in order to overcome high levels of wastage. Although much pollen is likely wasted, the quantities produced are more than enough to result in successful reproduction. Scientists have calculated that a single birch catkin produces around five and a half million pollen grains and a single rye floret generates some 50,000 grains. High levels of airborne pollen in summer are responsible for 17.6 million people suffering from hay fever in the USA alone.

Most wind-pollinated deciduous trees flower in early spring before their leaves come out. This means that pollen has a better chance of being intercepted by a stigma than by a leaf. Pollen rain from anthers onto stigmas located within the same flower would be very high, swamping them and preventing pollen from other individuals from landing. For this reason, many wind-pollinated trees have evolved to have the male and female sexes located in separate flowers. The average distance traveled by pollen from wind-pollinated trees ranges from tens to a few hundred yards (metres). That for wind-pollinated herbaceous plants is less, spanning a few yards at most. However, there are a few cases where the wind has carried pollen much farther. For example, southern beech (*Nothofagus*) pollen showed up in peat on Tristan da Cunha, a volcanic island in the south Atlantic, some 3,000 miles (48,280 km) from the nearest southern beech trees growing in South America.

1 *A solitary hairy-footed flower bee (*Anthophora plumipes*) uses its long (½inch/14mm) tongue to collect nectar.*

2 *An Anna's hummingbird (*Calypte anna*) feeding from a flower.*

POLLINATION BY ANIMALS

A surprisingly wide range of animals are involved with pollinating plants, including beetles, ants, flies, thrips, moths, butterflies, bees, wasps, birds, and bats. Beetles were probably among the earliest pollinators, having been abundant in the late Mesozoic, when gymnosperms were in their prime and angiosperms were on the rise. The present-day association of beetles with primitive woody angiosperms may reflect this early evolutionary path. Over time, the flowering plants evolved to make use of other animal visitors to their flowers and to maximize the efficiency of pollination. They did so by undergoing complex and intricate adaptations, through natural selection, to lure potential pollinators to them and then load them with pollen for their onward journey. Today, bees are the most prolific pollinators, pollinating around 80 percent of flowering plant species.

SWEET REWARDS

The primary reward offered by plants to animal pollinators is nectar. This carbohydrate snack, generated from readily available supplies of water, carbon dioxide and sunlight, is a much-needed food for visiting insects. They land on flowers, attracted by the prospect of a high-energy feed, and while there they inadvertently brush past pollen-loaded anthers. The pollen sticks to them, then, when they later land on another flower, it gets transferred to the awaiting stigmas. Pollination mostly results in cross-fertilization of flowers on different plants, but sometimes animal pollinators aid self-pollination of individual flowers.

Nectar is a sugar solution containing varying proportions of sucrose, fructose, and glucose. Sucrose can be broken down into equal amounts of fructose and glucose by the enzyme invertase, which is produced by yeast. Nectars rich in sucrose tend to be associated with flowers that have long nectaries (the glandular organ from which nectar is secreted). Such flowers are generally pollinated by long-tongued bees, butterflies, and moths or birds. Nectars that are rich in glucose and fructose are primarily associated with flowers that have exposed nectaries. These tend to be pollinated by short-tongued bees, flies, and tropical bats.

Naturalist Charles Darwin famously predicted the existence of the long-tongued moth *Xanthopan morganii* subspecies *praedicta* after observing the long nectary on specimens of the orchid *Angraecum sesquipedale*. "I have just received such a box full from Mr Bateman with the astounding *Angraecum sesquipedalian* [sic] with a nectary a foot long. Good heavens what insect can suck it," he wrote in 1862. After a few days spent observing his latest botanical acquisition, he concluded that: "in Madagascar there must be moths with proboscis [sic] capable of extension to a length of between ten and eleven inches." The moth exhibiting such characteristics was only discovered and named in 1907 after Darwin had died. And Darwin's theory was not proven to be correct until 1992, when images were taken of *X. morganii* transferring pollen between plants while feeding.

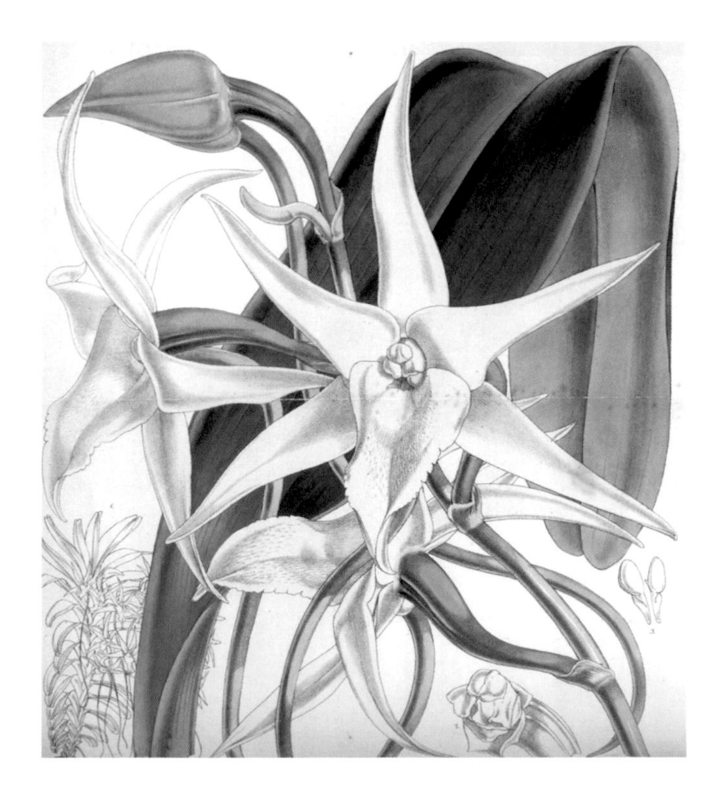

Charles Darwin predicted that the orchid Angraecum sesquipedale, *from Madagascar, would be pollinated by a particularly long-tongued moth.*

Flowers pollinated by birds, bats, and butterflies tend to produce dilute nectars of around 15–25 percent sugar content by weight, while nectar collected by bees tends to contain more than 50 percent sugar. Bees are special among insects in that they feed on both energy-giving nectar and protein-rich pollen, and therefore make more visits to flowers than other pollinators. Some plants have adapted to this need by providing abundant pollen and less nectar. Meanwhile, some 2,000 plant species provide a fatty oil instead of, or as well as, nectar. Secreted from organs called elaiophores, these attract highly specialized bees, which use the oil to feed their larvae and make their nest cell linings more water-resistant.

Much of the rich diversity of flowers seen on Earth today has resulted from plants adapting to maximize their chances of successful pollination by animal visitors. Plants engage shapes, colors, patterns, scents, and textures to guide potential pollinators to their flowers. Once the visitors have landed, characteristics such as tubular or trumpet-shaped flowers and symmetrical patterning help steer pollinators to the positions that will be most effective for pollination. Many flowers exhibit specific "guide marks," which are often only visible to humans under ultraviolet light but which show up clearly to insects. Sometimes, guide marks change color once pollination has been achieved, deterring subsequent visits from pollinators.

Some plants and pollinators have such a close relationship that the life cycles of each are entirely dependent on each other. Known as a mutualistic relationship, it is exemplified by the 800 or so species of tropical figs (*Ficus*), each of which has evolved a relationship with a particular species of fig wasp (Agaonidae). "Although we eat figs as fruits, they actually comprise a very specialized structure that is part inflorescence and part fruit," explains Simon Hiscock. "Called a syconium, when small it functions as an inflorescence full of little male and female flowers. When a female wasp, which has been flying around carrying pollen, gets the instinct to lay its eggs, it moves towards a tiny syconium that is in the flower stage, and it forces its way in through a little hole into the internal cavern."

Once inside, the female wasp deliberately pollinates some flowers, by placing pollen onto their stigmas, and oviposits her eggs into others. The ovules with eggs develop a gall-like structure instead of a seed; provided only a few ovules are affected within each fig, the benefits of the wasps' pollination services outweigh this loss to the plant. Most fig wasps lay their eggs in inner ovules, closest to the fig center, while most seeds develop in the outer ovules, near to the outer fig wall. The reason for this is because the offspring of fig wasps that develop in outer ovules are at risk of attack from parasitic wasps, which lay

1 *A daisy flower photographed under ultraviolet (UV) light. The yellow petals have lost their color, and nectar guides are now visible to the human eye. Bees, which have eyes that are sensitive to UV light, detect these nectar guides as prominent dark patches on the petals leading to the center of the flower.*

2 *A female fig wasp entering a ripening fig.*

their own eggs in outer ovules. The parasitic wasps attack from outside of the fruit, and kill the pollinator wasps' offspring.

Scientists believe that the parasitic wasps help to keep the number of eggs deposited by the pollinator wasps in check, to ensure the plant can grow enough seeds for its future success. When the pollinator grubs hatch out, the males emerge first and mate with females before they have even left their galls. The wingless males never leave the fig. "Their sole purpose is to mate with the females," says Simon. "Once they've done so, the males chop off the anthers of the male flowers of the fig and the females gather up the pollen and deliberately put it into pollen pouches in their thorax. The males then dig a hole in the fig syconium to let the females out. The males die inside the fig and females leave and start the life cycle all over again. The fig can't be pollinated by anything other than the fig wasp and the fig wasp can't complete its life cycle without the fig."

POLLINATION MAY FORM NEW SPECIES

Work by Kew Gardens' scientist Félix Forest has brought new evidence suggesting that pollination may drive the formation of new plant species, a process called speciation. He studied the evolution of plants belonging to the painted petal iris genus *Lapeirousia* in the Greater Cape of South Africa. *Lapeirousia* is a relatively small group of 27 species with diverse pollination systems; seven different pollination "syndromes" (suites of characteristics that identify a specific pollinator) have been reported in the group. They include two different "guilds" of long-proboscid fly pollinators from families of tangled-veined flies (Nemestrinidae) and horse flies (Tabanidae). By constructing a phylogenetic tree, showing relationships between different *Lapeirousia* species based on their DNA, Félix and his colleagues found that 17 shifts in pollinators had taken place during the *Lapeirousia's* evolution. This suggests the diversity in species within this genus might have been influenced by adaptations evolved to attract pollinators.

1

BEES ARE THE WORLD'S PRIMARY POLLINATORS

Plants exhibit diverse adaptations for attracting bees, but these have some common characteristics. Plants with yellow and blue or purple flowers with a three-dimensional form are favored by bees. And flowers with petals that are arranged in tubular or bell shapes tend to be pollinated by bees, as do ones that have bilateral symmetry (zygomorphy), such as monkshood, deadnettles, and louseworts. Bee-pollinated flowers also often have scents; for example, the grape hyacinth *(Muscari neglectum)* smells of plums, the Spanish broom (*Spartium junceum*) of honey, and the gorse (*Ulex europaeus)* of coconut. In turn, some bees scent-mark flowers they visit, warning subsequent bee visitors to avoid seeking out nectar from those flowers.

Orchids have the most outlandish adaptations to entice bee pollinators, which include making their flowers look and smell like female bees. The brown and yellow coloring, and slightly "furry" texture, of the yellow bee orchid (*Ophrys lutea*), for example, help to trick male *Andrena* bees into thinking they are approaching a female bee sitting on a yellow flower. This orchid does not offer any nectar reward. The plant simply deceives the male bee into thinking it is finding a mate, a trait known as "pseudocopulation." "The orchid is hijacking the uncontrollable innate response of the male bee in its desire to mate," explains Simon Hiscock, Professor of Botany and Director of the University of Bristol Botanic Garden. "It gives an olfactory cue in the form of a sex pheromone mimic, and a visual cue because the orchids have an iridescent patch on their large liplike petal, which resembles the wings of the female. When the male bee lands on a flower, it then gets a tactile cue because the furriness of the lip of the orchid makes it think it's on the furry abdomen of a female. This releases the final mating behavior."

The complexities of obtaining nectar make bees loyal to species that reward them well. Bees store the nectar they collect as honey, and feed it to their larvae along with the pollen they collect. The bees add the enzyme invertase to the nectar, breaking down the sucrose to glucose and fructose. Following water loss through evaporation, the sugar content of honey ends up being around 80 percent. Scientists have calculated that bees visit approximately 500 flowers during a single foraging trip lasting 25 minutes, and that producing a pound (about 450 g/16 ounces) of honey from the nectar of white clover flowers, which are small and numerous, takes approximately 17,330 foraging trips. Each pound of honey therefore represents food rewards from around 8.7 million flowers, requiring 7,221 hours of bee labor.

Some bees have learned to "rob" nectar from plants. For example, bumblebees with short tongues, such as *Bombus terrestris*, have been observed punching holes directly into the nectary of bell-shaped flowers on the strawberry tree (*Arbutus unedo*). This enables them to take nectar that would be simply beyond the reach of their tongues via the usual route. This is disadvantageous for the plant because the bee does not pollinate the plant. Moreover, "secondary robbing" sometimes takes place, when subsequent insect visitors also exploit the hole cut by the bee.

In recent years, bee numbers have been declining; the number of honeybees has fallen by 75 percent, for example. By pollinating crop plants, bees provide a valuable ecosystem service that is estimated to be worth $14 billion per year to the USA and £20 million ($30 million) to the UK. Without bees, food production systems around the world would collapse. Studies into the reasons behind bee declines have identified biodiversity loss as a possible cause. One study found that bees feeding on pollen from a single species had less healthy immune systems than those that ate pollen from a range of plants; a healthy immune system is required so that bees can sterilize food for their colony. Another study revealed that bees and the wild flowers that they rely on for food are declining in step.

1 *Some orchid flowers use deception to attract pollinators. The yellow bee orchid has evolved to look like a bee.*

PLANTS AND THEIR POLLINATORS

PLANT

JADE VINE

(Strongylodon macrobotrys)

POLLINATOR

BAT SPECIES

NATIVE TO

Rainforests of the Philippines

This evergreen vine grows vigorously in damp riverside locations, climbing up other plants to reach the light. Each of its long stems displays more than 90 striking pale green flowers. The flowers stand out at twilight, attracting bats to come and drink their nectar. As the bats hang upside down from the plant, their heads brush against the pollen. They deposit this pollen on the female parts of other plants they visit at a later time. Gardens such as Kew and Cornwall's Eden Project grow the jade vine as part of their living collections. In the absence of bats, they have to pollinate their plants by hand. The specimen in the Palm House at Kew Gardens flowers every two or three years.

PLANT

GIANT WATER LILY

(Victoria amazonica)

POLLINATOR

SCARAB BEETLES

(Cylocephata castaneal)

NATIVE TO

Slow-moving waterways across tropical South America

This vast water lily has leaves that grow to 8 feet (2.4 m) across. On the first evening a bloom opens; it is white and emits heat and the scent of pineapples, attracting scarab beetles to the plant. At this point, the flower is functionally female and receives pollen from plants visited earlier by the beetles. The flower closes, trapping the beetles inside, during which time they transfer pollen to the stigmas, prompting pollination. During the second day, the flower turns from female to male and the anthers start producing pollen. When the flower reopens on the second evening, it is pink and emits neither scent nor heat. The beetles, laden with pollen once more, now leave to seek out another white flower. Beetles require a high body temperature, so the heat they receive acts as a reward from the plant in exchange for their pollination services.

PLANT

ANGRAECUM CADETII

POLLINATOR

RASPY CRICKET

(Glomeremus orchidophilus)

NATIVE TO

Indian Ocean island of Réunion

Scientists only found out that crickets were capable of pollinating plants in 2010, when orchid researchers on Réunion filmed a raspy cricket carrying pollen on its head leaving a flower of *Angraecum cadetii.* They found a close match in size between the raspy cricket's head and *Angraecum cadetii*'s nectar-spur opening and deduced that the cricket was the plant's pollinator. Crickets usually eat vegetation, along with other insects, but may have evolved a taste for nectar due to a general lack of insects on Réunion. Claire Micheneau and Jacques Fournel, who made the discovery, scored a research double whammy, as the cricket caught on camera turned out to be a previously unknown species.

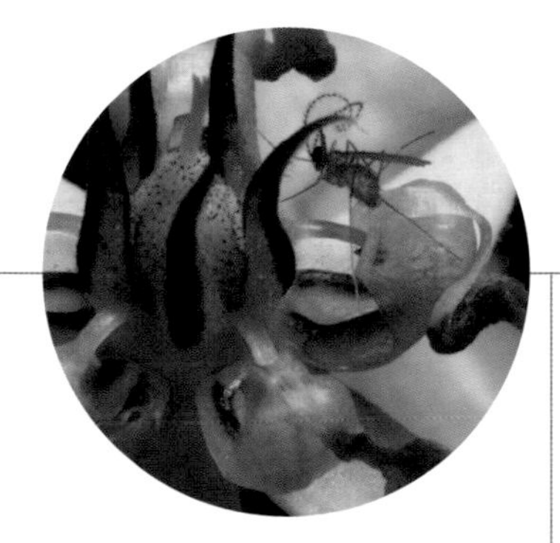

PLANT
THEOBROMA CACAO
POLLINATOR
MIDGES, THRIPS, AND APHIDS
NATIVE TO
Mexico, Central America, and northern parts of South America

The evergreen tropical tree *Theobroma cacao* has pods bearing seeds from which cocoa and chocolate are made. The indigenous people of Central America discovered its edible properties some 2,000 years ago. However, its method of pollination has taken longer for scientists to fathom. Midges from the genus *Forcipomyia* have long been considered a primary pollinator, but species from the genus *Dasyhelea* and *Stilobezzia* are also believed to contribute to pollination, along with some thrips and aphids. A study conducted in 2013 by scientists in Ghana found that 36 species of insect frequented cacao plantations but that more than half of them did not visit the *T. cacao* flowers. These observations, made over three years, supported the hypothesis that midges of the family Ceratopogonideae (which includes the *Forcipomyia* midges) were the most important pollinators.

PLANT
COFFEA SPECIES
POLLINATOR
HONEYBEES
NATIVE TO
Tropical Africa and Indian Ocean islands

Bees are usually rewarded for pollinating plants with a nectar or pollen snack. However, species from the *Coffea* genus and some citrus plants provide an additional caffeine shot in the nectar. This has been found to improve bee memories, helping them remember where to find the caffeine-giving flowers. In tests conducted by scientists at Newcastle University, Kew Gardens, and the University of Greenwich's Natural Resources Unit, honeybees feeding on a sugar solution containing caffeine were three times more likely to recall a flower's scent that those feeding on sugar alone. The caffeine benefits the plant and the pollinator; it improves the bees' ability to forage and makes them more faithful to the caffeine-giving plants.

PLANT
AXINAEA SPECIES
POLLINATOR
TANAGERS AND OTHER BIRDS
NATIVE TO
Central and South America

Tanagers are small- and medium-sized tropical birds. One of their favored foods is the stamens of species within the genus *Axinaea*. The plants have evolved to make use of the birds' liking for their male flower parts by developing bulbous bellows-like stamens that act like paper bags filled with air. These species exhibit clusters of partially opened flowers, against which the stamens clearly stand out. If a bird pulls at the stamens with its beak, it gets blasted by pollen that sticks to its feathers. When the bird goes on to visit other flowers of the same species, the pollen then rubs off and pollinates those flowers.

SEED BANK CHINA

SAVING CHINA'S DIVERSE FLORA

1 *The Germplasm Bank of Wild Species, in Kunming, Yunnan Province, China.*

2 *A first batch of UK seeds is handed over to GBOWS in 2008, for duplicating.*

3 *Inside GBOWS's cold seed store the temperature is a chilly -4°F (-20°C).*

NAME

Germplasm Bank of Wild Species, Kunming, Yunnan Province, China

NUMBER OF ACCESSIONS

China's Germplasm Bank of Wild Species (GBOWS) holds 65,067 seed accessions from 8,855 species (December 2014 figures). These represent around a third of China's native species of seed plants.

WHEN FOUNDED

GBOWS began operating in 2007, having initially been proposed in 1999 by the distinguished botanist Wu Zhengyi. It was officially opened in 2009.

FOCUS OF THE COLLECTION

GBOWS's overarching focus is to conserve wild plant species, by collecting and banking seeds. However, it also preserves plant tissue, using *in vitro* micropropagation methods, and has a DNA bank. Some animal tissue and microbial collections are also held at GBOWS.

WHY IT IS NEEDED

China has 31,500 plant species, of which half are endemic (not found anywhere else). It also contains four of the world's 34 "global biodiversity hotspots." One of these hotspots lies in the mountains of southwest China, where GBOWS is located, an area considered to have the most endemic-rich temperate flora in the world. In all, China has an eighth of the world's floral species growing within its boundaries. It is also home to a fifth of the world's population and its fastest-growing economy. The China Plant Specialists Group of IUCN's Species Survival Commission has listed 4,408 of the nation's species as endangered, around 15 percent of China's vascular plants.

WHO FUNDS IT

GBOWS is a research division of the Kunming Institute of Botany, Chinese Academy of Sciences. It is funded by the national government of China.

WHERE SEEDS ARE STORED

The Germplasm Bank facility has two wings. The South Wing houses the seed bank and the North Wing hosts research laboratories and administrative departments. GBOWS is situated within Kunming Botanic Garden, a 109 acre (44 hectare) site established in 1938. Some 5,000 plant species and cultivars are preserved in 13 specialist garden divisions. The garden attracts 500,000 visitors annually.

CURRENT RESEARCH

A major research project, the Barcoding Chinese Plants Project, started in 2009. It uses particular short and standard DNA regions from a plant's genome to identify the species. The long-term aim is to barcode some 6,000 species of plants in China, including the seed collection within the GBOWS facility, and to establish a reference library of Chinese plants. More than 60 scientists from 22 research institutes and universities are involved in this ongoing project.

1 *Researchers collect seeds at an altitude of 14,750 feet (4,500 m) in China's Sichuan province.*

2 *These seeds being bagged up are from the rare rowan* Sorbus pohuashanensis.

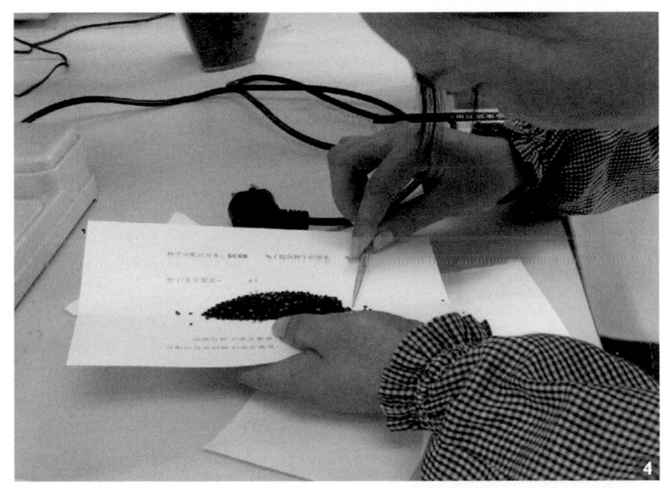

The ultimate goal is to develop an intelligent flora, called iFlora. This will incorporate DNA barcodes, diagnostic morphological plant characteristics, digital images, and geo-referenced collections for classifying and monitoring plant diversity. "In the next five years we hope to identify 80 percent of plant genera and 25 percent of species of China," explains Dezhu Li, the leader of GBOWS. "It's a very ambitious project."

SEEDS WITH A STORY

Scientists estimate that more than 5,000 plant species are used regularly as sources of traditional Chinese medicine (TCM). *Psammosilene tunicoides* is a rare plant endemic to southwest China. The roots of the plant are used in TCM to treat rheumatism and to enhance blood circulation. Recent scientific studies have also found the roots to contain chemicals with anti-fungal properties. Wild populations are severely threatened from habitat degradation and overexploitation. GBOWS's scientists have made eight collections, comprising 12,000 seeds, to ensure future generations can also benefit from this useful plant.

---

3 *Researchers clean seeds in the laboratory.*

4 *Each collection must be carefully counted before storage.*

5 *A flower of* Davidia involucrata. *Ernest Wilson gathered seeds of this highly sought-after plant in China for Kew Gardens in 1901.*

PLANTS AND SEEDS FROM ANTARCTICA AND THE ARCTIC

Antarctic and Arctic flora experience some of the most extreme conditions on Earth, with temperatures dipping to –40°F (–40°C) at times. Just two flowering plants exist on the Antarctic continent, while the Arctic is home to around 1,000 species. Adaptations include maintaining a small size, growing thick hairs, and having flowers that follow the sun to maximize use of its warmth. Levels of flowering and setting seed are generally low; reproduction by cloning predominates. Here is a selection of species that thrive in the world's polar regions.

1 *The aptly named Antarctic hair grass (*Deschampsia antarctica*).*

2 *The continent's only other plant species, the Antarctic pearlwort (*Colobanthus quitensis*).*

ANTARCTICA

DESCHAMPSIA ANTARCTICA

The Antarctic hair grass grows in maritime areas of the Antarctic Peninsula, a chain of 6,500 feet (2,000 m) high mountains, as far south as Refuge Islet. Tolerant to freezing, it grows to around an inch (25 mm) high in patches that are not normally larger than 270 feet^2 (82 m^2). Rapid climate change has affected the Antarctic Peninsula over the past 50 years, resulting in a total increase in mean annual air temperature of around 5°F (2.5°C)—several times higher than the global mean temperature rise. The Antarctic hair grass has become increasingly widespread as a result, as nitrogen released from thawing soils has provided it with additional plant "food." However, genetic diversity is generally low among populations; research suggests that vegetative reproduction and self-fertilization are the predominant drivers of the colonization of new sites.

COLOBANTHUS QUITENSIS

The Antarctic pearlwort is also limited to the Antarctic Peninsula, growing as far south as Neny Island. With the Antarctic summer only 12 weeks long, the pearlwort has to flower and set seed quickly. Growing to around ¼ inch (6 mm) high in dense cushions, it has tiny leaves and masses of small, cup-shaped yellow flowers on long stalks. Like the Antarctic hair grass, its distribution has widened with rising temperatures. Seeds of both Antarctica's native flowering plants are stored at the MSBP for safekeeping. Tim Rich, a plant taxonomist at the National Museum of Wales, collected them from the South Shetland Islands, located just off the tip of the Antarctic Peninsula, while he was on vacation there.

THE ARCTIC

PAPAVER RADICATUM

The yellow flowers of *Papaver radicatum*, the Arctic poppy, are heliotropic; that is, they turn to face the sun as it moves across the sky with the Earth's rotation. This helps the flowers to trap the sun's heat to the extent that, on a clear, sunny day, the poppy can attain a temperature 18°F (10°C) above that of the surrounding air. Making the most of available warmth is important for setting seeds within the short Arctic summer season. In experiments where Arctic poppy flowers were prevented from tracking the sun, the quantity and the quality of their seeds was reduced.

SALIX ARCTICA

The Arctic willow only attains a height of 4 inches (10 cm), but grows a horizontal trunk that can make a single plant as wide as its more southerly relatives are tall. Growing horizontally protects it from the biting Arctic wind. Plants are unisexual (male or female) and both have dense clusters of upright catkins. Flowering takes place from June to August. The tree reproduces by seed and vegetatively by rooting at the nodes of stems.

The young leaves, stems, and buds of Arctic willow are an important source of food for wildlife, including ptarmigan, muskox, and reindeer. The tree is also useful to the Inuit, who use the roots, leaves, and bark to make baskets and clothing. The Yakuts, who inhabit eastern Siberia, make a tea substitute called chai-talak from the willow. In regions where there are no other sources of wood, the Arctic willow is used as a fuel.

PULSATILLA PATENS

This small, mauve-flowered plant grows across the Arctic tundra and in northern latitudes around the world. Its stems and leaf undersides are covered with white hairs, which protect it from the cold. It has no petals but instead exhibits petal-like sepals. It is early to flower, sometimes doing so even before the winter snows melt. The naturalist and author John Muir wrote of it that: "Instead of remaining in the ground waiting for the warm weather and companions, this admirable plant seemed to be in haste to rise and cheer the desolate landscape."

P. patens's common name of windflower aptly evokes the way its seeds are dispersed. Once fertilization has taken place, the stem grows longer and the flower falls away to leave wispy seed heads. Its achenes (small single-seeded fruits that do not open to release their seed) are carried on the wind; once they land, they twist their way into the soil.

1 *Arctic poppies (*Papaver radicatum*) track the sun as it crosses the sky.*

2 *The windflower's mauve flowers herald the arrival of spring after the long Arctic winter.*

3 *The trunk of the Arctic willow (*Salix arctica*) grows horizontally, to protect it from the freezing wind.*

4 *Wispy seed heads of the windflower (*Pulsatilla patens*).*

2

3

4

YEW

(TAXUS SP.)

ENGLISH YEW Taxus baccata

PACIFIC YEW Taxus brevifolia

HIMALAYAN YEW Taxus wallichiana

GENUS Taxus

FAMILY Taxaceae

SEED SIZE Seeds are enclosed in a fleshy aril up to ⅜ inches (10 mm) long

TYPE OF DISPERSAL Animal. Arils are eaten by birds and squirrels, which excrete the seeds.

SEED STORAGE TYPE Orthodox

COMPOSITION Data not available

Yew trees grow across the northern hemisphere in Europe, Asia, North Africa, and North America. Often surviving for millennia, they have long been associated with healing and spirituality. In Britain, the common yew (*Taxus baccata*) is often seen growing in churchyards and is associated by Christians with the resurrection of Jesus Christ. Some of these trees pre-date Christianity, however, and churchyards often have links to the very earliest forms of worship, suggesting that yews may also have been of spiritual significance in pre-Christian Britain. Scotland's Fortingall Yew, which stands in Fortingall churchyard, is reputed to be 5,000 years old.

In the 1960s, the US National Cancer Institute distilled the drug paclitaxel (Taxol®) from the bark of the Pacific Yew Tree, *Taxus brevifolia*; it was subsequently found to be present in all 11 species of yew. Later clinical trials proved Taxol's efficacy as a chemotherapy drug. Considered to be one of the most important natural products to emerge in the twentieth century, in this form paclitaxel has revolutionized treatments for breast and ovarian cancers. Taxol was first synthesized in the laboratory in 1993; however, clippings are still used as part of the process of making the drug. Scientists estimate that it takes three tons of *Taxus* leaves to produce 2 lb (900 g) of paclitaxel. Globally, some 2,000 lb (900 kg) of the drug are required each year.

Ongoing use of the Himalayan yew (*Taxus wallichiana*) for medicinal uses, in Western as well as Ayurvedic, Tibetan, and Unani medical traditions, are threatening its survival; it is listed as endangered by the IUCN. However, China's Germplasm Bank of Wild Species holds around 30,000 *T. wallichiana* seeds and the MSBP also holds *T. wallichiana* seeds within its collections. These, along with tissue cultures, could be used in future to shore up wild populations of the species and to assist with its cultivation. Meanwhile, recent studies have revealed that hazelnuts and several species of fungi also yield Taxol, which may enable scientists to produce the much-sought-after drug in commercial quantities in future.

1 *A seed of the common yew (*Taxus baccata*).*

2 *The seeds are contained in a fleshy aril, which attracts birds.*

3 *Yews have long been associated with healing.*

CHAPTER 4

DISPERSAL TAKES SEEDS TO NEW PASTURES

THE DIVERSE WAYS IN WHICH PLANTS SPREAD THEIR SEEDS

The success of an individual plant species depends to a great extent on its ability to distribute its seeds to places where they can germinate and thrive. Some plants rely on gravity or ballistic mechanisms to propel their seeds a short distance from the parent plant; others have evolved ways of attracting animals to eat them or carry them away on their fur; plants in blustery areas have adaptations that help the wind to carry their seeds away; while others rely on rivers or the sea for dispersal, the seeds sometimes traveling vast distances on ocean currents. Humans, too, have helped some plant species achieve great success, and by the same means have contributed to the demise of others, by transferring plants and seeds around the world.

PARACHUTING IN

Many seeds have adaptations that aid dispersal by a particular method. Take the common dandelion (*Taraxacum officinale*), for example. Each flower produces a seed that is enclosed in an achene. Each achene, of which there can be 180 in a single plant, is topped by a pappus of filaments. When these are dry, they form a parachute shape that helps the achene and its enclosed seed become airborne. This adaptation makes seed dispersal by the wind highly effective, with models of seed dispersal indicating that dandelion seeds can travel several miles.

HIT AND MISS

Plants dispersed by animals often offer a reward in the form of food. However, unlike with pollination, where plants and animals frequently display co-evolutionary traits that benefit both parties, seed dispersal by animals is a much less exact science. "In forests, some birds like to sit in clearings because they can keep an eye out for predators and sing to attract mates," explains Ken Thompson, a plant ecologist at the University of Sheffield's Department of Animal and Plant Sciences. "If those birds drop seeds they've eaten into the clearing, it's a bonus for the plant because its seeds get moved out of the shade of the mother plant into some sunlight. So that can be a very good thing. However, seeds also quite often get dispersed to places where they are not likely to thrive. It's a very chancy business."

NEAR AND FAR

The variation in distances traveled by seeds is immense. Plants that rely on self-propulsion to disperse their seeds, such as the Himalayan balsam (*Impatiens glandulifera*), achieve distances from the mother plant of less than 33 feet (10 m). Dispersal by ants, too, covers relatively short distances. Seeds carried by the wind or by larger animals, meanwhile, can travel much farther; African forest hornbills have home ranges that can exceed 10,000 acres (more than 4,000 hectares), and some individuals have been observed to travel as far as 180 miles (290 km) over a two- to three-month period. Water, too, is effective for long-distance seed dispersal; ocean currents are known to carry seeds hundreds of miles from their point of origin.

1 *Dandelion (*Taraxacum officinale*) seeds being dispersed by the wind. Models indicate that dandelion seeds can travel several miles.*

2 *A pappus of filaments helps dandelion achenes and their enclosed seeds become airborne. Here, water droplets are contained by the pappus.*

2
1

THE CHALLENGES OF STUDYING SEED DISPERSAL

Scientists use various approaches when trying to monitor seed dispersal. One is to find a plant of interest and then mark the seeds. Some researchers have marked them with fluorescent paint and then tried to find them at night with an ultraviolet light that makes them glow. Others have sprayed seeds with a radioactive isotope and then attempted to locate them by detecting the radioactivity. "Somebody years ago did a study where they looked at dandelion dispersal," explains Ken Thompson of the University of Sheffield. "There was a patch of dandelions releasing seeds on a windy day and each student in a class of undergraduates was given the job of chasing one seed until it landed."

Seed traps can be placed at locations spreading away from a plant, but seeds that have traveled a long distance can be quite difficult to find. Scientists at the UK's Centre for Ecology and Hydrology experimented with trapping seeds of heather (*Calluna vulgaris*) and bell heather (*Erica cinerea*). The seeds of both are numerous and very small; in a single season, one *Calluna* plant can generate more than seven million seeds, each measuring less than 1/32 inch (0.6 mm) across. Using an increasing density of traps with distance from the plant, the researchers recorded the wind-dispersed seeds up to 260 feet (80 m) away. This is the widest area over which a dispersal curve has been measured for non-tree species. For wind-dispersed seeds, an alternative approach is to try and model seed dispersal. This involves taking a particular seed and measuring its aerodynamic behavior in a wind tunnel. "You measure the wind and the turbulence and eddies out in the real world and then you do some fancy math and try to figure out how far your seeds are likely to go," says Thompson.

Finding seeds dispersed by animals is one of the hardest challenges for seed scientists. Tracking devices, such as GPS transmitters, can help. Scientists at the Smithsonian Tropical Research Institute in Panama attached GPS transmitting backpacks to six wild toucans to monitor their distribution of seeds from the common Panamanian nutmeg tree (*Virola nobilis*). The researchers worked out the time taken for the toucans' guts to process nutmeg seeds and then plotted the birds' journeys. They calculated that the toucans were dropping nutmeg seeds an average of 472 feet (144 m) from the parent tree.

With the advent of DNA technology, scientists have analyzed dispersed seeds to find their genetic fingerprint and then gone looking for the plant that shed them. One study that analyzed dispersed seeds of the mahaleb cherry (*Prunus mahaleb*) tree found that no more than five trees contributed seeds to a particular landscape patch and that up to 62 percent of seeds were distributed within 50 feet (15 m) of the source tree. Long-distance dispersal events occurred infrequently. "All seed dispersal studies are hard to do, whatever the approach," admits Thompson. "That's why we tend not to know what happens to most seeds once they leave the parent plant."

EFFECTS OF CLIMATE CHANGE

Difficult as it is to monitor where seeds go after being shed, understanding the mechanisms by which seeds are dispersed is becoming more important in light of the negative impacts of human activities on the world's flora. Scientists estimate that a fifth of the world's estimated 380,000 flowering plant species face extinction. Climate change and biodiversity loss are placing great pressure on both plant and animal populations. If animal seed dispersers become rare or go extinct, the plant populations that rely on them could decline.

The methods plants use to disperse their seeds may also have an impact on their ability to survive in a warmer climate. Scientists have found that Arctic and alpine plant species with limited capacity for seed dispersal are likely to lose more of their genetic diversity in a warmer climate than plants that use animals or the wind to distribute their seeds farther afield.

One potential loser is the glacier crowfoot (*Ranunculus glacialis*), which lives on high mountaintops in Scandinavia and southern Europe, and has little gene flow between separate populations. Unable to disperse its seeds far, this could mean that it loses entire populations if conditions become too warm. On the other hand, plants such as the wind-dispersed dwarf birch (*Betula nana*) may have a better chance of long-term survival, as they may be able to shift their ranges in line with changing climatic conditions.

HOW SEED SIZE MATTERS

Seeds have masses spanning more than 11 orders of magnitude, from the tiny specks of orchid seeds to the 44 lb (20 kg) seed of the coco de mer (*Lodoicea maldivica*) palm. Scientists who analyzed seeds from around 11,500 species, along with the locations in which they grew, found that seeds are, on average, larger closer to the tropics and smaller farther from the equator. They recorded a 320-fold decline in seed mass between the equator and 60° latitude (see diagram). At the edge of the tropics, they detected a sevenfold drop in mean seed mass. The changes were most closely related to changes in plant growth form and type of vegetation (trees, shrubs, and so on).

While it might seem likely that seeds with a smaller mass would be dispersed farther than larger ones, the height of a plant is, in fact, a more important determinant of how far its seeds will travel. Scientists reviewing 148 studies from around the world studied 200 plant species with varied seed-dispersal methods, and concluded that seed dispersal distance is more closely correlated with plant height than with seed mass. This refuted a widely held perception that small-seeded species should disperse better than large-seeded species, trading off seed mass (a greater seed mass equating to a larger number of nutrients provided by the parent plant) against dispersal capacity.

SEED MASS REDUCES WITH DISTANCE FROM THE EQUATOR

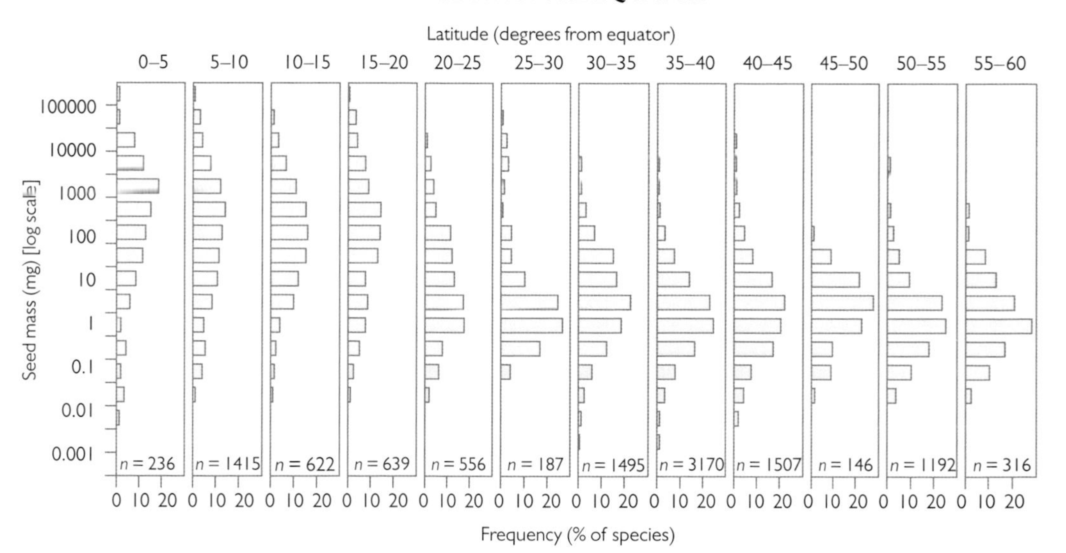

*The seeds of glacier crowfoot (*Ranunculus glacialis*) do not disperse far, which could limit its chances of survival in a warmer world.*

ANIMALS DISPERSE SEEDS FAR AND WIDE

There are four main types of zoochory, the term used to describe seed dispersal by animals. The first is through birds or rodents hoarding seeds for later retrieval but never eating them, allowing the seeds to germinate. The second is by birds, mammals, and even fish eating seeds and then defecating or regurgitating them at locations away from the parent plant. The third is where ants carry seeds underground to eat a food body on the seed surface, known as the "elaisome," and then abandon them. And the fourth type is via adaptation such as barbs, hooks, and stickiness attaching seeds to animal fur or feathers.

BURIED TREASURES

Hoarded seeds usually end up in holes in trees or in the ground. Only if the animal forgets where it buried them, or dies, do the seeds get the opportunity to germinate. However, hoarding is a remarkably effective form of seed dispersal. An example comes from the "Wilderness" at Monks Wood in the UK. This 10 acre- (4 hectare-) field, bounded on three sides by woodland, has not been sown with crops for several decades. When scientists studied seed dispersal here, they found that oak seeds dispersed by wood pigeons, jays, and gray squirrels were distributed widely across the field, while the distribution of wind-blown ash (*Fraxinus*) seeds was limited largely to the edges of the field, close to the boundary with the woodland.

FEED ON SEEDS

Seeds that are hoarded do not necessarily exhibit any obvious adaptations. However, most seeds dispersed by vertebrates offer a nutritious pulp—fruit—as a food reward. Mammal-dispersed seeds are often scented and dark-colored, reflecting their dispersers' predominantly nocturnal habits. Wild Elmleaf blackberry (*Rubus ulmifolius*) seeds, for example, have been observed being eaten by red foxes (*Vulpes vulpes*) and stone martins (*Martes foina*). Seeds dispersed by birds tend to be sheathed in brightly colored fruits that do not smell. The white berries of the European mistletoe (*Viscum album*) are particularly attractive to the European mistle thrush (*Turdus viscivorus*), for example. *V. album* is parasitic, growing on trees such as apple, lime, and poplar. Its seeds are dispersed when birds wipe the seed-containing pulp off their beaks onto the branches of a potential host.

ANTS HELP OUT

More than 80 plant families rely on ants to disperse their seeds. Myrmecochory, as this mutualistic relationship is known, is particularly important within Europe and North America's temperate deciduous forests and in dry shrubland plant communities in Australia and South Africa. As with mammals and birds, ants often receive a food reward in return for the service they provide. This is the elaisome, which is rich in lipids and proteins. Attracted by a chemical in the fleshy body, the ants carry seeds into their nests. Here they feed off the elaisome before discarding the intact seed on the nest's refuse pile. Although ants do not disperse seeds far (research studies indicate a mean global distance of around three feet), by burying the seeds the ants remove them from the surface, where they may otherwise get eaten by rodents or damaged by fire. In experiments using the scrambled eggs plant (*Corydalis aurea*), ant-planted seeds were demonstrated to produce 90 percent more seedlings than those planted by hand. Some ants bury the seeds to eat later, in the same way that squirrels hoard food. This tends to be the case in desert environments, where ants are very common.

A STICKY BUSINESS

The final kind of dispersal by animals involves adaptations that help the seeds stick to fur or feathers. In this case, the plant does not offer a reward; rather, adaptations such as hooks and barbs cause the seeds to adhere to animals as they pass. Unlimited by the "gut-passage" time, seeds carried on the outside of animals often travel farther than those that are eaten. Examples of seeds dispersed by attachment include goosegrass (*Galium aparine*), barbed goatgrass (*Aegilops triuncialis*), and beggarticks (*Bidens pilosa*). Paul Smith, former head of Kew Gardens' Millennium Seed Bank Partnership, and now Secretary General of Botanic Gardens Conservation International, encountered plants from the *Uncarina* genus with particularly sticky adhesive burrs on its fruits when he was in Madagascar. The species probably evolved to be dispersed by larger mammals that no longer exist on the island. "The plant is so sticky that it can trap small mammals; it sometimes even kills them because they can't get away," says Paul. "It's an amazing-looking thing. I saw a snake skeleton hanging on one plant."

HUMANS HELP SPREAD ALIEN INVADERS

Humans often inadvertently pick up and carry seeds as they walk through low vegetation. In Australia, they have even been accused of helping to spread invasive species in this way. Scientists studied how hikers might play a role in dispersing seeds from five invasive weeds: bidgee-widgee (*Acaena novae-zelandiae*), sheep's sorrel (*Rumex acetosella*), sweet vernal grass (*Anthoxanthum odoratum*), cocksfoot grass (*Dactylis glomerata*), and red fescue grass (*Festuca rubra*). They found that in one season alone in Kosciuszko National Park, New South Wales, visitors could carry 1.9 million seeds around the park on their socks and 2.4 million seeds on their trousers. The type of clothing had a great impact on the number of seeds dispersed; for example, seeds attached to socks better than to pants. Meanwhile, spiny bidgee-widgee seeds were more likely to attach to clothing than smoother cocksfoot seeds.

*A South American harvester ant (*Pheidole aberrans*) carries a grass seed to its nest in Santa Fe, Argentina.*

ELEPHANTS AS PRIMARY SEED DISPERSERS

In many tropical rainforests, where the wind cannot penetrate and there is great competition for space, it makes sense for plants to engage the services of passing animals to disperse their seeds. It is also advantageous to develop sizeable seeds containing plentiful nutrients, as without much light for energy-creating photosynthesis at ground level, seedlings need as large a supply of food as they can get. Thus the perfect dispersal agent for these plants is an animal big enough to carry large seeds some distance from the parent plant to a location suitable for germination.

Elephants fit the bill perfectly. Slaves to eating, they need to munch their way through around 330 lb (150 kg) of vegetation on a daily basis. And they are also partial to large-seeded fruits such as mango (*Mangifera indica*), jackfruit (*Artocarpus integer*), and wood apple (*Limonia acidissima*), a tennis-ball-sized hard fruit native to Bangladesh, Pakistan, Sri Lanka and India. Moreover, they do not chew the seeds much before swallowing them and their digestive processes have only a mild effect.

"If several elephants come to a tree, they may find many fruits on the ground and they can eat a large number of them at once," explains Ahimsa Campos-Arceiz, Associate Professor of the School of Geography at the UK University of Nottingham's Malaysia campus. "They can move long distances and they will disperse seeds for several days in different places. The elephants essentially plant the seeds, most of which come out in a viable condition, in a little mound of poo. It's a great place to start to germinate. So elephants are very effective seed dispersers."

SIZE MATTERS

The trouble is, there are no longer many of them around. The spread of farming, the expansion of urban areas, and a high demand for ivory have caused elephant numbers to fall and their habitats to shrink. Prior to the arrival of humans, these animals were at the top of the food chain, escaping predation, so they evolved few mechanisms to react against a high level of mortality. Today, the Sumatran elephant (*Elephas maximus sumatranus*) is critically endangered; the Indian elephant (*Elephas maximus indicus*) and Sri Lankan elephant (*Elephas maximus maximus*) are endangered; and the African elephant (*Loxodonta africana*) is vulnerable. Globally, elephants are known to disperse as many as 500 plant species. So what happens to plants that have evolved to use elephants as seed dispersers when these animals disappear?

"In Africa, we think of an elephant as a combination of a bulldozer and a gardener," says Ahimsa. "Although elephants go around making a mess, breaking off branches and trampling plants, this creates more habitats for plants and other animals to inhabit. When an elephant goes into a rainforest and breaks branches, pulls off leaves and defecates, it creates crevices for little lizards and insects to live in and it makes plants accessible to animals on the ground, as well as to those that live in the canopy. Meanwhile, its dung provides a rich resource of nutrients. So you have a very dynamic system. When elephants disappear, all this heterogeneity disappears and you get a much more homogeneous, dull set of conditions."

SIGNS OF CHANGE

At present there is little hard data to show what happens to individual plant species that lose their primary seed disperser. But there is plentiful anecdotal evidence. The elephant apple (*Dillenia indica*) is a good example. It grows in India, where elephants are known to eat and disperse the seeds. When Ahimsa came across the plant in Myanmar, however, it was described as water-dispersed and was mainly found growing by rivers. As elephant populations had dwindled in the country, the plant's main seed-dispersal mechanism had become less effective. Where the elephant apple grew by rivers, its seeds were simply carried downstream where new populations grew beside the water. "When megafauna disappears, seed dispersal becomes very localized, shrinking from kilometers to tens of meters [miles to yards]," Ahimsa explains. "Then the whole dynamic of the population is compromised. This brings into question the long-term survival of the population and, eventually, the entire species."

ENDANGERED SPECIES THREATEN SEED DISPERSAL

According to WWF's 2014 Living Planet report, global populations of mammals, birds, reptiles, amphibians, and fish halved between 1970 and 2010. Many large

*An African elephant (*Loxidonta africana*) reaches up into the branches of a tree to feed on seed pods, in Manu Pools National Park, Zimbabwe.*

mammals that disperse seeds, such as chimpanzees, rhinoceros and tigers, are among the most endangered species. So is there a way in which plants can survive after their primary disperser has gone? In the case of the mango, when no elephants or other suitable large mammals are around, its seeds are sometimes dispersed by primates, such as macaques, which might grab a mango, eat a little of the fruit and carry it a few tens or hundreds of yards (meters) away from where they found it. However, dispersal by primates greatly limits the area over which new mango plants can grow.

Ahimsa had thought that tapirs might play a similar ecological role to elephants and could therefore be a good alternative dispersal agent for mangos. Tapirs are herbivores that grow to around the size of a large pig. However, they, too, are no match for the seed-dispersing qualities of elephants. "Unfortunately, we found there were bigger differences than size alone," Ahimsa explains. "Elephants will quickly swallow mangos or jackfruits but tapirs just chew and chew. They also have much longer digestion times. So tapirs kill most of the seeds they eat through mastication or digestion. And if they can separate the seeds from the fruit they will just eat the fruit. So we found that for plants with big seeds, the tapir was a very bad disperser."

REWILDING AND COEXISTENCE

The idea of "rewilding," where large animals that have disappeared from a habitat are reintroduced, has been gaining ground in recent years but is still a challenging prospect. Elephants, for example, are not just large; they can cause great damage to humans, buildings, and crops. Ahimsa believes that for rewilding to work, you need to have areas that are dedicated solely to farming, where crops are protected, and which are no-go areas for the animals. Then you need to designate areas purely for conservation. And alongside both, you need places where humans and elephants can coexist but where people have the materials and expertise they need to protect their assets, with compensation mechanisms in place in case the animals cause damage.

One project, in Sri Lanka, is having some success in getting this balance right. Prithiviraj Fernando, Chairman, trustee, and scientist at the Centre for Conservation and Research in Sri Lanka, and Research Associate of the USA's Smithsonian Institution, has been assisting communities by providing them with materials and skills to build protective electric fences. In exchange for managing and maintaining the fencing, the farmers are given compensation if an elephant damages their crops or buildings. Ensuring that people and their livelihoods are protected could be one way to help humans and megafauna coexist without conflict in the same space. In the long term, it might even offer the best way to shore up wild populations of plants that evolved to depend on large animals for their survival.

FLOATING DOWN RIVERS AND ACROSS THE HIGH SEAS

The term hydrochory is used to describe seed dispersal by water, whether in fresh or seawater. Enabling seeds to be dispersed over long distances, it plays a role in expanding the range of species, and in connecting local plant communities and populations. It may also allow species to migrate in the event of climate change. More negatively, it is implicated in the spread of non-native invasive species, which become quickly established and have a negative impact on local ecosystems.

There are three types of hydrochory: nautochory, where seeds are dispersed by water currents at the surface; bythisochory, where seeds are dispersed by water currents on the bottom of a channel or on the ground following heavy rain; and ombrochory, where seeds are dispersed by rain falling on plant stems and leaves. A fourth category, zoohydrochory, describes situations where dispersal takes place jointly by animals and water. This includes dispersal by humans and their watercraft, fish, beavers, and turtles.

FLOATING DOWNSTREAM

Scientists have found a relationship between the capacity of species' seeds to float and the frequency of species along rivers. This suggests that rivers act as corridors for plant distribution and may help to maintain regional biodiversity. However, if seeds were continually dispersed downstream and eventually washed out to sea, we might expect that the upper reaches of streams would be depleted in river-dispersed species while the lower reaches would be concomitantly enriched in those same species, the logical corollary being the eventual extinction of species in the upper reaches.

For many plant species, this downstream drift does not seem to occur. Recent genetic analyses suggest that dispersal between catchments, including that taking seeds upstream, occurs in many plant species. This movement, mediated by the wind, birds, and fish, acts to counteract the constant downstream flow of the water. A study of seed dispersal by pacu, very large fish related to piranhas, reveals the extent to which fish can play a role in moving seeds around aquatic environments.

Scientists found that two species of pacu, dwelling in the rivers and floodwaters of South America, disperse seeds from around 35 percent of local plant species.

BATH TOYS HIGHLIGHT EFFECTIVENESS OF DISPERSAL BY OCEAN CURRENTS

The extraordinary distances that objects can travel on ocean currents was highlighted by an accident at sea in 1992. During a stormy January night, while traveling between Hong Kong, China, and Tacoma, Washington, the ship *Ever Laurel* shed twelve 40 feet (12 m) containers into the Pacific Ocean. One of the containers held Friendly Floatee plastic bath toys; when it subsequently broke open, its cargo of 28,800 green frogs, red beavers, yellow ducks, and blue turtles was released into the sea. Some toys journeyed north while others went south.

It took ten months for the first toys to wash up in Alaska. Three years later, they began appearing in Japan, other parts of North America, and Hawaii. Oceanographers deduced that, after initially heading north, they had joined the subpolar gyre, a current that flows anti-clockwise in the Bering Sea. Some toys had escaped the current and traveled farther north, becoming frozen in Arctic ice. They traveled slowly eastward before reaching the North Atlantic and starting to journey south. The toys that traveled directly south from the Pacific eventually landed in Australia, Indonesia, and South America.

Pacu are among some 200 species of fish that move into temporarily flooded areas during the wet season. They eat fruit that falls into the water, later depositing the seeds onto floodplains. Once floodwaters have receded, the seeds are able to germinate and grow into seedlings. In this way, seeds are dispersed upstream as well as downstream.

Coconut seeds are dispersed by ocean currents. Here, a germinated seed floats on the sea surface close to shore.

SEAFARING SEEDS

Most plant species are capable of being dispersed by freshwater, but only a few are adapted for dispersal in saltwater. The coconut (*Cocos nucifera*) is well-known for having seeds dispersed by ocean currents. Scientists have demonstrated that coconut seeds can germinate even after floating in seawater for 110 days.

The term "drift seeds" is used to refer to a range of parts from plants' reproductive organs that wash up on shores around the world. Some are obviously well adapted for ocean travel. The innermost layer of the pericarp (part of the fruit surrounding the seeds) of the blister pod (*Sacoglottis amazonica*) has air-filled cavities to help it float. Other drift seeds use adaptations such as corklike coats or spongy cotyledons to reduce their density in water. An impermeable outer coat prevents tissues from absorbing seawater. Embryos protected in this way stay dormant during transportation; some can remain viable for many years. The longest-recorded distance traveled by a drift seed—namely, by a Mary's bean (*Merremia discoidesperma*) from the Marshall Islands to Norway—is 17,400 miles (28,000 km).

LONG-DISTANCE DISCOVERIES

Drift seeds from tropical plants found on temperate shores have long fascinated beachcombers as well as scientists. Charles Darwin was among the early naturalists who investigated their properties. He carried out experiments on 87 species of vegetables and other plants by immersing them in bottles of brine for various lengths of time. He concluded that ocean dispersal was possible for 14 percent of the seeds he tested in this way, proposing that seeds of 14 percent of the plants of "any country might be floated by sea currents during 28 days across 924 miles (1,490 km) of sea, to another country, and when stranded, if blown to a favorable spot… would germinate."

Today, sharp-eyed beachcombers in northwestern Europe, including Britain and Ireland, can sometimes pick up drift seeds such as the nickar nut (*Caesalpinia bonduc*) from the tropical Americas, including the West Indies and Florida; the sea bean (*Entada gigas*) from the tropical Americas and the West Indies; and the mangrove palm (*Nypa fruticans*) from Asia and the Pacific regions. From the tropical plant's point of view, having seeds dispersed as far as temperate climates is of little use, as they are unlikely to germinate on arrival at their final destination. However, green-fingered beachcombers often relish the challenge of coaxing long-distance drift seeds into life.

HITCHING A RIDE WITH THE WIND

An array of structural adaptations enables plants to capitalize on wind power to transport their seeds. This form of seed dispersal is termed anemochory. The dandelion "clocks" and spinning sycamore seeds loved by children represent the two primary methods that anemochorous plants employ to carry their seeds to pastures new: having light seeds that float on the wind and aerodynamic seeds that glide or spin their way slowly to the ground. The majority of adaptations slow the rate of the seed's fall from the parent plant, enabling the wind to carry it laterally.

Plants with wind-blown seeds are often pioneer plants; they can grow in harsh conditions and poor soils. They are therefore mostly found in open country, in steppes, open heaths or downs, deserts, sand dunes, and along the boundaries of forests, riverbanks, and roadsides. Wind dispersal is common among the floras of Australia, New Zealand, and North America. Around half the plants of the Mediterranean garrigue scrubland have windblown seeds, and 70 percent of those in Alaska are anemochorous. Wind dispersal of seeds is ancient; winged seeds, for example, date back to the Devonian, not long after seeds themselves evolved.

HAIRS HELP SEEDS' AIRBORNE JOURNEYS

Plants that have lightweight seeds with fine "hairs" attached to them are found all over the world. There are many variations on how hairs are arranged to aid a seed's journey through the air. One plant with "fuzzy"

*When seeds of the sycamore (*Acer pseudoplanatus*) are shed, they spin to the ground.*

MODELING SEED DISPERSAL

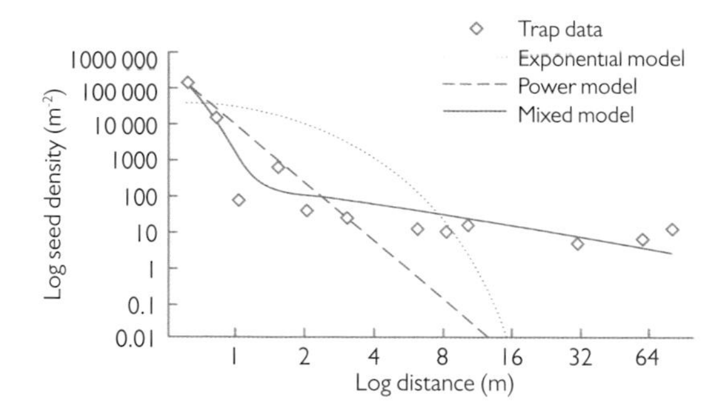

*Comparison of seed-trap data with predictions from three empirical models for common heather (*Calluna vulgaris*) seeds, trapped along a 260 feet (80 m) transect in Dorset, UK.*

hair-covered or "comose" seeds is cotton (*Gossypium hirsutum*), the fibers of which are used to make textiles. Several plants of the Compositae family, to which the common dandelion belongs, engage in the parachute method of seed carrying. And other lightweight seeds, such as those of old man's beard (*Clematis vitalba*) and the windflower (*Pulsatilla* sp.), have long, feathery extensions, like kite tails, which help keep them airborne.

WINGED FLIGHT

Winged seeds, meanwhile, rotate or glide to the ground. They are usually found on tree species, as these provide sufficient height for the seed to travel a reasonable distance horizontally before reaching the earth. The gyrating effect of sycamore (*Acer pseudoplatanus*), ash (*Fraxinus*), and hornbeam (*Carpinus*) seeds relies on air resistance, the weight of the seed and the shape of its wing. Other plants produce seeds with wings that glide rather than spin. The most well-known of these is the Javan cucumber, *Alsomitra macrocarpa*, a type of climbing gourd. It has seeds that can be 6 inches (15 cm) across but are only paper thin. Lifted by the gentlest of breezes from football-sized pods, the seeds travel hundreds of yards (meters) across the Indonesian rainforest.

GERMINATION WARFARE

Seeds of the sycamore and the Javan cucumber have both inspired aircraft designs. Defense contractor Lockheed Martin emulated the natural engineering of the sycamore in the design of its 12 inch- (30.5 cm-) wide unmanned drone, the Samarai monocopter; the name derives from "samara," the generic term used to describe winged seeds and fruits.

Meanwhile, the performance of the Javan cucumber's seeds inspired Igo Etrich and Franz Xaver Wels to build gliders with a similar wing shape. In 1906, Etrich produced the Taube, a fixed-wing sports plane, using a similar design. The German word *Taube* means "dove," the symbol of peace, but the plane went on to become the first mass-produced military plane. It was the first plane to drop explosives, in 1911, during the Italian–Turkish war.

Japanese engineers Akira Azuma and Yoshinori Okuno also studied the aerodynamics of *Alsomitra* seeds. They found that the seeds descend at an angle of 12 degrees, which means they fall 16 inches (40.6 cm) each second, compared to the 40 inches (1.2 m) per second observed in many rotating winged seeds.

MEASURING WIND DISPERSAL

Traditionally, studies that have plotted the "number of windblown seeds" against "distance traveled" have displayed a leptokurtic curve. This is one where the highest number of seeds falls close to the parent, while the number reaching farther distances gradually declines. Direct measurements of wind dispersal in the field often show that most seeds are dispersed very short distances, only rarely traveling distances greater than tens of yards (meters). However, given the difficulties of measuring wind dispersal, especially at greater distances from the parent plant, some caution has to be taken when interpreting results. For example, the study of dispersal of heather and bell heather conducted by the Centre for Ecology and Hydrology (see page 108) found that the majority of seeds fell close to the parent plant, but after a few yards (meters), there was little further decline in density. This indicated that methods resulting in a leptokurtic curve may be underestimating the long-distance component of dispersed seeds.

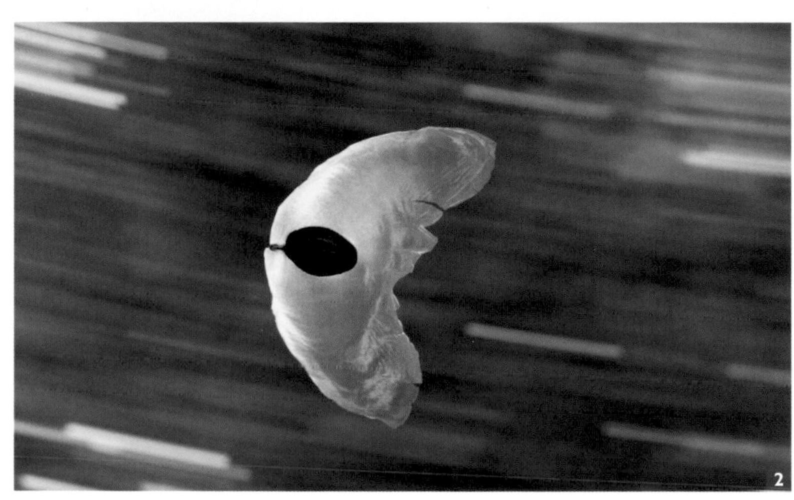

1 *A seed from the Spinning-top conebush (*Leucadendron rubrum*), suspended from a pappus of hairs, is carried away from its parent plant by the wind.*

2 *A winged seed from the climbing gourd (*Alsomitra macrocarpa*) glides through the tropical rainforest in West Papua, New Guinea, Indonesia.*

Measuring wind dispersal accurately is important because scientists believe that the long-distance component may govern the rate at which plant populations expand. Studies suggest that stronger winds, such as those unleashed in rare storms and initial uplift by thermals or turbulent eddies, are both important in promoting long-distance dispersal of seeds; with uplifted seeds possibly traveling tens of miles. Evidence of both the pioneering nature of wind-dispersed seeds and their ability to travel long distances comes from studying freshly created habitats. Scientists found that species with more conspicuous adaptations for wind dispersal were more likely to colonize avalanche debris in Japan, landscapes altered by the eruption of Mount St. Helens in the USA, glacial forelands in Switzerland, and spoil heaps in northern England.

CREATING LIFE AFTER DEATH

The word "tumbleweed" refers to plants that employ the wind to scatter their seeds once they have died. In this extraordinary adaptation, the dried plant disengages from its roots and is then bowled along by the wind, either scattering seeds as it goes or waiting until it comes to rest before shedding them. If the conditions are suitable, new plants will then grow in this location.

Tumbleweeds are common in desert or steppe environments. Plants that form them include: pigweed (*Amaranthus retroflexus*), a weed that grows in the western USA; Russian thistle (*Salsola sp.*); and the Australian spinifex grasses.

DISPERSAL BY GRAVITY AND BALLISTIC PROPULSION

Dispersal by gravity (barochory) or self-propulsion (autochory) are relatively minor methods of seed dispersal. With these mechanisms, seeds either fall directly to the ground or are explosively propelled away from the parent plant. The seedpod of the common poppy (*Papaver rhoeas*) has holes in the top, through which seeds are shaken out like pepper from a cruet when the stems bend over. The marigold, meanwhile, uses gravity, combined with an adaptation to the shape of its seeds, to help ensure its offspring's success. The seeds are elongated with feather-like arrows at one end, which helps the tips of the seeds tilt downwards. This shape makes it easier for the seeds to penetrate the soil surface, aiding their chances of burial.

Most plants that use ballistic methods of dispersal are relatively small in stature, although a few trees also employ this mechanism. One plant that has used this method with great success is the small pink-flowered filaree or common stork's bill (*Erodium cicutarium*). It is thought to be native to Europe, North Africa, a stretch from West Asia to eastern Russia, and the Indian subcontinent. However, following introductions in the eighteenth century, it is now naturalized throughout southern and central Australia, New Zealand, wider parts of Asia, and throughout North, South, and Central America.

The key to its reproductive success lies in its awns (slender bristles attached to its seeds, much like those commonly seen on barley or wheat ears). While moist on the plant, the awns are held straight, but after flowering, as the fruits of the plant dry, the awns coil and form a spring mechanism that abruptly throws the seeds 18 inches (45 cm) or so from the plant. Once free from the parent, the awns coil and uncoil with changes in relative humidity, physically drilling the seeds into the soil. Backward-facing hairs on the awn force the seeds to move in one direction, hence drilling continues even when the awn uncoils.

SEEDS BURST FORTH

Other methods of ballistic projection are fruits that burst suddenly, often prompted by heating from the sun, as exemplified by species of *Ulex*, *Euphorbia*, and *Geranium*. Seedpods of the yellow-flowered heathland plant, gorse (*Ulex europaeus*), are known for bursting open in summer with a cracking sound. The species *Euphorbia boetica*, a Mediterranean perennial, has been shown to shoot its seeds explosively as far as 26 feet (8 m). One of the strangest self-dispersing plants is the squirting cucumber (*Ecballium elaterium*), the fruit of which fills up with slimy juice as it ripens. The stalk acts like a stopper until, eventually, the buildup of hydraulic pressure is so great that the fluid bursts out, spraying the seeds several yards (meters) away from the plant.

*Macrophotograph of the ripe seed head of a common poppy (*Papaver rhoeas*). The blue-black seeds are being released from the window-like pores in the top of the dried-out capsule.*

HOW SEEDS ARE DISPERSED AROUND THE WORLD

2 FISH AND TERRAPINS DISPERSE EELGRASS SEEDS

Research conducted in Chesapeake Bay, USA reveals that marine fish and terrapins may, alongside birds, help disperse the seeds of eelgrass, a flowering plant that grows on the seabed.

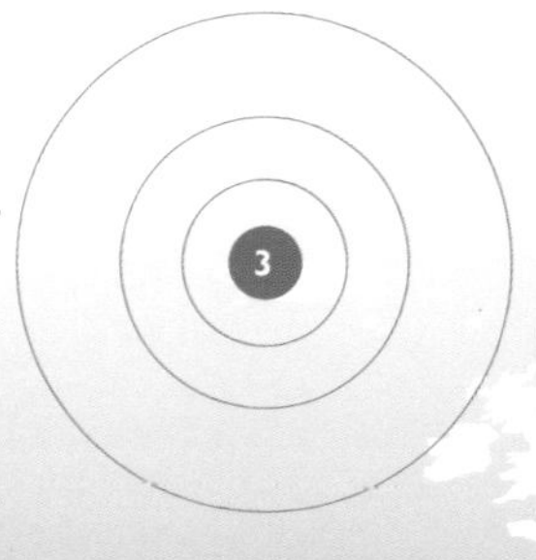

1 THE HAIRY PRAIRIE SEED DISPERSER

The bison (*Bison bison*) is an important seed-dispersing animal within central North America's tallgrass prairies. Scientists who analyzed dung and hair samples for seeds found 70 plant species represented within dung samples and 76 species represented in hair samples. In each case, around half the seeds present were from grasses.

3 NEW PLANT INHABITANTS ARRIVE BY SEA

Almost 80 percent of the numerous plant species that colonized the volcanic island of Surtsey, near Iceland, in the decade after it emerged in 1963 were carried there by ocean currents. However, only a quarter of these had apparent morphological adaptations for dispersal by water.

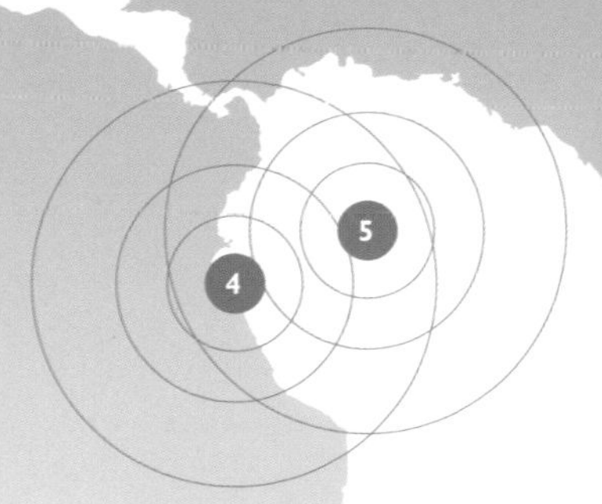

4 LONG-DISTANCE MYSTERY

Scientists have long debated the extent to which ocean dispersal was responsible for the coconut's existence across southwest Asia and Melanesia as well as on the Pacific coast of America. Recent DNA evidence and modeling studies indicate that, while dispersal by sea is likely to account for its distribution over shorter distances, particularly historically between emerging coral atolls, the coconut was, in fact, brought by seafarers from the Philippines to Ecuador some 2,250 years ago.

5 ANIMAL DISPERSERS RULE IN RAINFORESTS

Dispersal by animals and birds dominates in tropical rainforests. Some animals may be particularly dominant within a particular ecosystem. For example, scientists found that red howler monkeys dispersed seeds from 137 species in Central Amazonian rainforests during a 25-month study period, making them highly effective seed dispersers.

Most seed dispersal methods occur within most ecosystems, but some are more prevalent than others under particular geographical conditions. This diagram highlights some of the important dispersal mechanisms at work in different environments around the world.

6 ROLLING DISPERSERS

In autumn in Russia's Ural Mountains, mature plants of Russian thistle (*Salsola sp.*) dry out and detach from their roots. They then begin rolling in the wind, shedding seeds as they go. A single plant typically has 250,000 seeds. Dropped seeds germinate when temperatures warm in summer.

7 MIXED METHODS

Of 128 tree species growing in a tropical semi-evergreen forest in northeastern India, 78 percent were found to be dispersed by animals and 22 percent by wind or gravity.

8 RARE ELEPHANTS UNDERPIN CONGO FLORA

The forest elephant, a subspecies of the savanna-dwelling African elephant, may disperse more seeds in the Congo than any other animal. Scientists found that forest elephants ate seeds from more than 96 plant species, distributing them as far as 35 miles from the parent plant. The Congo's elephants have declined by 80 percent in the past 50 years; if they become extinct, the Congo's forest ecosystems may undergo drastic changes.

9 BIRDS AND BATS VITAL FOR AUSTRALIA'S SEED DISPERSAL

Twenty-five species of birds (including the topknot pigeon, Lewin's honeyeater, and the bowerbirds) and three species of bats (two flying foxes and a tube-nosed bat) disperse more than two-thirds of the plant species that exist in Australia's subtropical rainforests.

SEED BANK USA

THE SEED BANK KEEPING NEW YORK CITY GREEN

NAME

Greenbelt Native Plant Center,
New York

NUMBER OF ACCESSIONS:

The seed bank at the Greenbelt Native Plant Center (GNPC) currently holds 3,384 accessions but is very much an active collection. "It's like a conveyor belt," explains Director Edward Toth. "We're constantly drawing seeds down and restocking. We're probably never going to have more than around 5,000 accessions. The whole point is to use the seed and get it back out in the environment. It's kind of a revolving door, where the numbers stay fairly stable."

WHEN FOUNDED

The GNPC grew out of a small restoration project undertaken in Prospect Park, Brooklyn, sister park to New York's Central Park. In 1990, plants were needed to restore a small woodland ecosystem in Prospect Park, but at that time no native plant nurseries existed that could provide plants for the project. The GNPC was founded to grow the required stock; over three years, it grew close to a half a million plants. "The facility we were using for the nursery was one of the last surviving farm properties in the city," explains Edward, who was Land Manager of Prospect Park at the time. "It had been slated for development, but through some lucky happenstances we were able to rescue it and keep it in the public sector. When the project drew to an end, my boss and I appealed to the Commissioner [of the Department of Parks and Recreation] to turn the venture into a citywide resource. He agreed, so I left Prospect Park and took up my current role."

Seedlings growing up in a glasshouse at the Greenbelt Native Plant Center in New York.

FOCUS OF THE COLLECTION

The primary focus of the GNPC's seedbank is to provide locally sourced seed to produce plants for the center's nursery. The plants grown are used almost exclusively in New York City on public projects by the municipal Department of Parks and Recreation and other agencies to manage public lands. Over time, GNPC's mission has widened, so it also contributes seeds for the Seeds of Success (SOS) Program and the Mid-Atlantic Regional Seed Bank (MARS-B). The former initiative is a federal program run out of the Bureau of Land Management, which aims to provide a national seed bank for the USA. The latter is a regionally focused, project-based venture established by the GNPC. One MARS-B project that the center contributes to is an initiative to gather seeds from all ash (*Fraxinus*) species across the mid-Atlantic states. This is in response to attacks by the emerald ash borer, a destructive non-native beetle. Specific funding allows the GNPC to focus for three years on banking seeds from ash trees within the state of New York.

To meet the needs of larger projects, where direct sowing of seed is the only viable economic approach, the center also produces native, locally sourced bulk seed. Around an acre and a half of the center's land is therefore devoted to growing "foundation" seed. The aim, in time, is for growers to take the foundation seed and to use it to make large-scale increases in seed volumes. They will then sell the bulk seed back to New York City for use in its land-management projects. "We have foundation seed for about 35 species at this point, and we're looking to have it for upwards of 100 species," explains Edward. "Those 100 species will be components of seven or eight seed mixes that we will make available to land managers in the city. A lot of them are meadow plants; there are some for wet meadows and some for dry meadows. We also have coastal grasslands, so we

1 *Herbarium "vouchers"; dried plant specimens that the GNPC uses to help identify unknown species.*

2 *Oak seedlings at various stages of growth.*

3 *This machine helps to clean the seeds.*

4 *Here, seeds are undergoing an initial drying process.*

have maritime grassland mixes, and we're working on some woodland mixes. We're also developing an urban mix, comprising colonizing species found in depleted urban soils. We see a use for a mix like this in vacant lots and brownfield sites."

WHY IT IS NEEDED

Many habitats are native to the city of New York, from coastal salt marshes to upland forests. When new urban developments, or natural disasters such as 2012's devastating Hurricane Sandy, encroach upon remaining natural areas, the best response is to try to restore habitats using plants that originally grew in that location. At any one time, the GNPC has between 350,000 and 500,000 plants growing in its nursery for such purposes. The key to effective plant conservation is the use of locally sourced, genetically diverse seed stock. A key mission of the GNPC is to promote this localized approach across the region and the USA as a whole. GNPC, MARS-B, and SOS all function to further this goal. "Seed is a critical natural resource that needs wise management," says Edward.

WHO FUNDS IT

The center is wholly funded by the city of New York and operates under the municipality's Department of Parks and Recreation. With between 20 and 30 staff, including two full-time seed collectors, it is the largest and best-funded native plant nursery at municipal level in the USA. Grants support new initiatives that explore and expand the mission of the center.

WHERE SEEDS ARE STORED

Located on Staten Island and covering 13 acres (5.2 hectares), the GNPC comprises six greenhouses, five acres (two hectares) of irrigated outdoor nursery, and the seed bank. Seeds are stored in a floral cooler (a kind of refrigerator used by florists) with an industrial humidifier installed. This maintains the seed at temperatures of between 60°F and 65°F (15°C and 18°C) at relative humidity of between 5 and 15 percent. Under these conditions, seeds will stay viable for decades. Everything is accessioned in much the same way as a museum collection. Whoever collects the seed gathers information about both the site and the conditions under which they gather the seed. That information is fed into a database and remains associated with the individual collection to which it relates.

CURRENT RESEARCH

The GNPC does not get involved in implementing projects; it simply grows the plants for them. "We try to be responsive to whatever plants are needed," explains Edward. "For example, the city was pretty hard-hit by Hurricane Sandy so there's a need right now for plants for coastal restoration projects, such as dune stabilization and replanting maritime grasslands."

1

2

SEEDS WITH A STORY

Less than ten days before Hurricane Sandy struck in 2012, the GNPC's head seed collector made a collection of seeds from wild populations of American beach grass *(Ammophila breviligulata)* growing on the Rockaway Peninsula of New York's Long Island. This species holds migrating dunes in place, so it is widely used in dune stabilization programs. However, because wild populations are hard to find, much of the *Ammophila* available for such projects is a cultivar from the US Department of Agriculture. As cultivars have no genetic variability, they are not very resilient to disease or climate change. When Hurricane Sandy struck, the Rockaway Peninsula was one of the hardest-hit areas; the storm completely wiped out the dune plant communities, including the populations of *A. breviligulata*. "It was very serendipitous that we had seeds from those wild populations in our seed bank, as they are now being used to help restore the dune vegetation," says Edward.

1 *Larger seeds are placed in cloth bags for banking.*

2 *Smaller seeds are stored in envelopes.*

3 *Accessions are carefully recorded in the seed card catalogue.*

4 *New shoots begin to emerge from recently sown seeds.*

PLANTS AND SEEDS FROM THE WORLD'S ISLANDS

Cut off from major landmasses by the sea, islands often experience the rapid development of new plant species and they are frequently home to high numbers of endemic species. Ensuring the survival of unique island floras is a high priority for conservation organizations and botanic gardens, as they are often at high risk from invasive alien species. Here are some particularly rare island species.

ST. HELENA, SOUTH ATLANTIC

MELISSIA BEGONIIFOLIA

Officially listed by the International Union for Conservation of Nature (IUCN) as critically endangered in 2003, *Melissia begoniifolia*, the St. Helena Boxwood, is now considered to be extinct on its native island home. This branching shrub, with a scent akin to smelly feet or tobacco, was once sufficiently widespread to have a hill, Boxwood Hill, named after it. However, its numbers dwindled to nothing over the years. It had long been considered extinct when a single population of the shrub was found in 1998, hugging a stony slope on the southwest of the island. This plant eventually succumbed to drought and infestations by pests.

There is a glimmer of hope that the plant could yet survive in the wild as some seedlings have emerged from seeds that must have been dormant in the soil. However, the "extinct" label will remain unless these grow to become adult flowering plants that can reproduce and expand the population once more. If this fails, the St. Helena Boxwood has a final chance for survival through a reintroduction program. Seeds collected from the last remaining plant before it died have been grown up at Kew Gardens and have been used to generate a further 13,000 seeds. As well as being stored for safekeeping at Kew's MSBP, seeds are being sent in batches back to St. Helena. These are being grown into small plants for establishing in areas of the island where *M. begoniifolia* was previously widespread.

1 *The survival of* Melissia begoniifolia *hangs in the balance on the Atlantic island of St. Helena.*

2 *The rocky Hawaiian island of Nihoa is the only site where* Amaranthus brownii *grows.*

2

HAWAIIAN ISLANDS, NORTH PACIFIC

AMARANTHUS BROWNII

Located in the middle of the Pacific Ocean, more than 2,000 miles (3,220 km) from the nearest continent, the Hawaiian islands are the most isolated in the world. This isolation has contributed to the islands' plants' high level of endemism; around 89 percent of native plants growing there are unique to the islands. The main threats to these species are competition from introduced species and fires.

Amaranthus brownii grows only on the uninhabited island of Nihoa, within the northwestern Hawaiian islands. A needle-leafed, green-flowered annual herb, which only grows on rocky outcrops at altitudes between 390 and 700 feet (120 and 215 m), it is the rarest plant on the island. Specimens of it were initially collected in 1923 by botanist Edward Leonard Caum. On a specimen voucher from this trip, now held in the Herbarium at Kew Gardens, he noted that the plant was "scattered, all over island".

By the early 1980s, however, only two populations of 35 plants were known to exist, and subsequent expeditions have failed to spot any specimens (primarily because surveys take place in summer but the plant only grows in winter). The IUCN lists it as critically endangered. Although botanists believe it still exists, it is threatened by competition from non-native pigweed (*Portulaca oleracea*), growing numbers of vegetation-scoffing gray bird grasshoppers (*Schistocerca nitens*), fire, and hybridization with other *Amaranthus* species. So far, efforts to propagate *A. brownii* from collected seeds have been unsuccessful.

MONTSERRAT, CARIBBEAN

RONDELETIA BUXIFOLIA

The small island of Montserrat lies within the Leeward Islands of the Lesser Antilles. Since the mid-1990s, life there has been dominated by repeated eruptions of the Soufrière Hills volcano. In 1997, a major eruption devastated the southern half of the island, burying the capital Plymouth. As well as killing 19 people, it wiped out large swathes of the island's vegetation.

Before the eruption, *Rondeletia buxifolia*, the pribby, which has glossy dark-green leaves and pale orange flowers, was only known from dried specimens of the plant gathered in 1979. But surveys conducted on the island in 2006 found the species to be growing in the wild. Its distribution was limited to an area less than 11 square miles (17 km^2), however, mostly located outside of the island's protected forest area. A new botanic garden on Montserrat, set up by the Montserrat National Trust with assistance from Kew to replace the original one in Plymouth, propagated the pribby. It also planted a demonstration hedge of the species to assess its potential for replacing non-native species in hedges across the island. Seeds of the species are now held for safekeeping at the MSBP.

GREEK ISLANDS, MEDITERRANEAN

AETHIONEMA RETSINA

This critically endangered member of the mustard family (Cruciferae) is known to inhabit just two Greek islands: Skiros and Skiropoula. Three populations live close to the seashore on vertical limestone cliffs between 30 and 650 feet (10 and 200 m) above sea level. A shrubby perennial, which forms cushions, *A. retsina's* fleshy appearance is an ecological adaptation to salt spray. Scientists are very interested to learn of its chromosomes and how they differ from agriculturally important members of the same family, such as cabbage, canola and mustard. It is possible that the salt- and drought-tolerant properties exhibited by *A. retsina* could be bred into related crops.

SEYCHELLES, INDIAN OCEAN

LODOICEA MALDIVICA

The towering double coconut palm *Lodoicea maldivica* has the largest seeds of any plant. In the wild, the trees are limited to two populations on the islands of Praslin and Curieuse in the Seychelles. The palms take between 25 and 50 years to grow to maturity, and the fruit can take two years to germinate. The suggestive double-lobed shape of the seeds, resembling the curves of a female bottom, gave rise to sailors' tales that they had aphrodisiac powers. Highly sought after today, the seeds are at risk from poachers, and trade in double coconut seeds is closely controlled.

1 *The landscape of Montserrat has been shaped by repeated volcanic eruptions.*

2 Rondeletia buxifolia, *the pribby, was presumed to be extinct until it was discovered growing within a small area on Montserrat in 2006. Seeds are now held at the MSBP, safeguarding its future.*

3 *The curvaceous seeds of the giant coconut palm (*Lodoicea maldivica*) have beguiled visitors to the Seychelles for centuries.*

MONGONGO

(SCHINZIOPHYTON RAUTENENII)

GENUS Schinziophyton

FAMILY Euphorbiaceae

SEED SIZE Up to 1 ¼ inch (30 mm)

TYPE OF DISPERSAL Animal (elephant and others)

SEED STORAGE TYPE Orthodox

COMPOSITION Oil: approx. 57%; Protein: approx. 26%

The mongongo is a large deciduous tree that grows in savanna woodlands in a belt across subtropical southern Africa. Its nuts, contained inside a very hard kernel within a velvety egg-shaped fruit, are an important source of nutrition. Containing protein, calcium, magnesium, iron, and vitamin E, they are eaten daily by subsistence communities across the region. The nuts have a taste similar to an almond or cashew, and also yield oil rich in linoleic acid, which is used to make skincare products.

The seeds are known to be dispersed by elephants, porcupines, ostriches, and kudu (a kind of antelope). In the case of elephants, only the fruit is digested; the nut and kernel remain intact upon defecation and are sometimes collected by locals from piles of dung, saving them the trouble of picking the fruit from the trees. When seed specialists from Kew's MSBP talked to communities in Botswana that use the mongongo, they found that wild populations were becoming depleted, possibly due to the loss of the plant's seed dispersers, and that locals found it hard to germinate the seeds and cultivate the trees.

Moctar Sacande, the MSBP's Research Leader, Diversity and Livelihoods, spent two years in the laboratory working out how best to germinate mongongo seeds. Eventually, he found that chipping the seed coats and pre-treating the seeds with a smoke solution, to replicate natural wildfire events, broke the seed's dormancy and resulted in successful germination. Initially, the MSBP's partner organization in Botswana, the National Tree Seed Center, previously had only been able to grow 15 trees from 100 seeds, but equipped with Moctar's germination protocols they were then able to grow 87 trees from the same number of seeds. "We managed to set up all the correct growing conditions in Botswana and now local communities in Tsetseng are growing thousands of mongongo seedlings," he says.

1 *A mongongo tree (*Schinziophyton rautanenii*) laden with fruit in Zambia.*

2 *Seeds of the mongongo.*

3 *Inside a mongongo seed. The hard outer coat is clearly visible.*

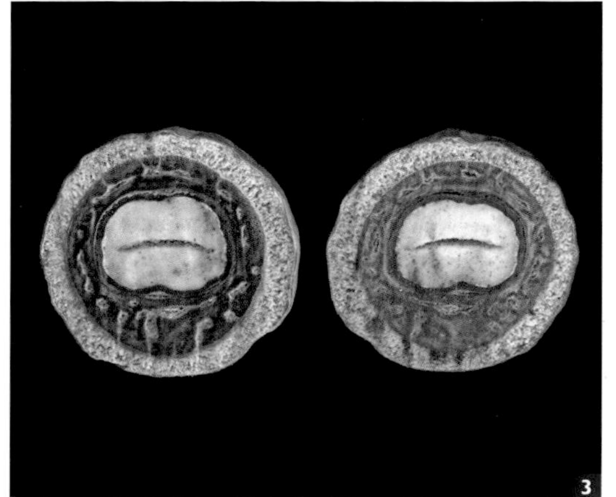

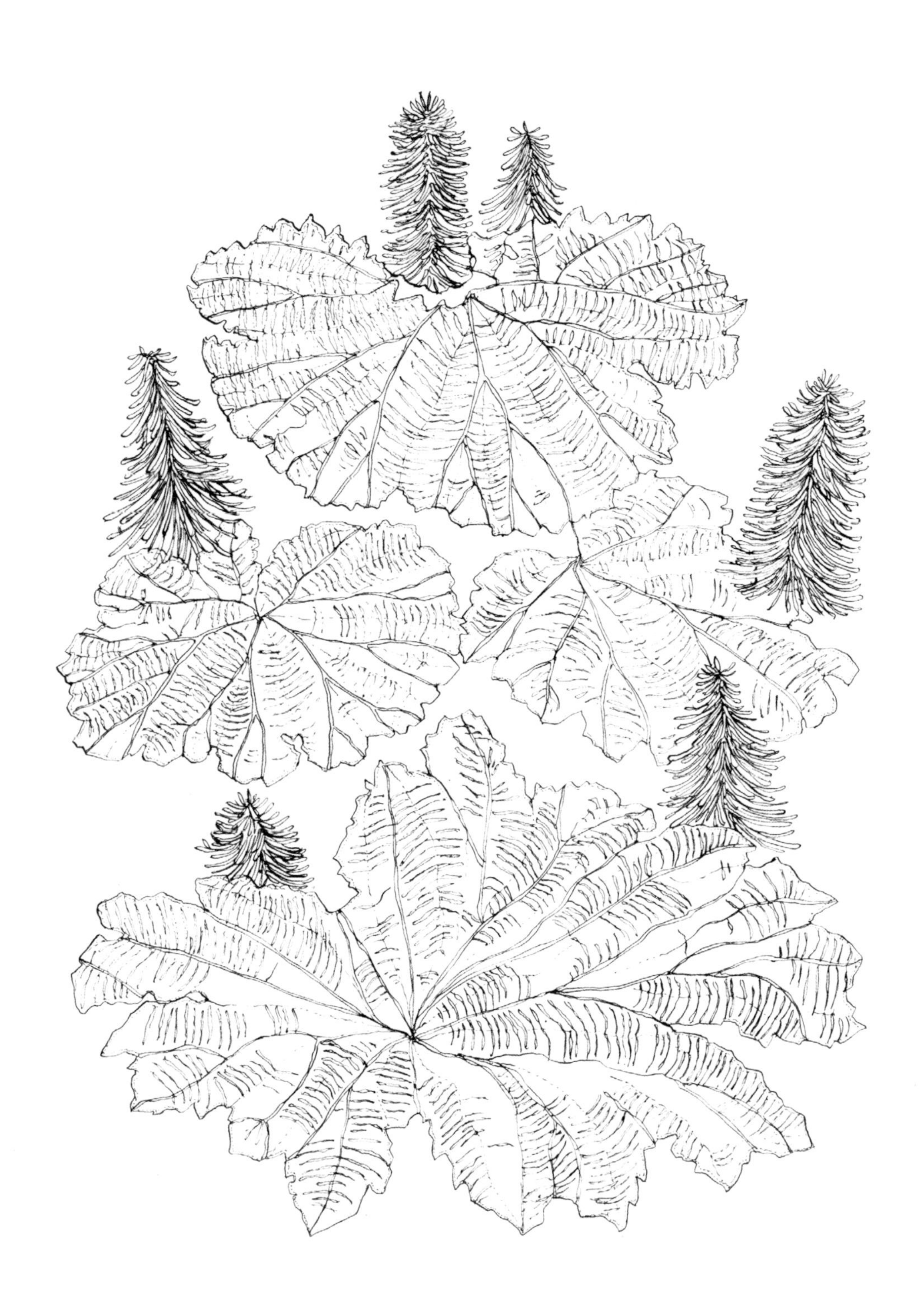

CHAPTER 5

GERMINATION BRINGS PLANTS BACK TO LIFE

THE TRICKS PLANTS USE TO SURVIVE

As all gardeners know, if you want plants to thrive you need to ensure the conditions in which you place them mimic those of their natural environments. There is no point in planting the sun- and moisture-loving Chilean rhubarb (*Gunnera manicata*) in complete shade in a well-drained bed, for example, or it won't have the huge leaves for which it is renowned. Similarly, you cannot expect a rhododendron from the acid soil of the Himalayas to grow splendidly on the lime-rich chalk soils of southern England.

SEED DORMANCY

Growing plants from seeds requires even greater skill. Even if you provide the conditions of light, temperature, and humidity favored by a particular plant, you may not be successful in coaxing its seeds to germinate, because many seeds have built-in dormancy. This enables them, over time, to sense signals that indicate the conditions for germination are becoming favorable for seedling establishment and to release the dormancy so the seeds can germinate. Only horticulturalists who emulate this array of signals will have a good chance of germinating seeds successfully.

Around 70 percent of plant species have some form of seed dormancy, varying according to habitat. Scientists have found that 85 percent of seeds from hot deserts, temperate deciduous forests, steppes, scrublands, and cold deserts exhibit dormancy once they reach maturity. Meanwhile, 60 percent of seeds from tropical rainforest plants and 50 percent of those from tropical semi-evergreen forests are non-dormant when mature. In temperate areas, dormancy prevents seeds from germinating during periods such as cold or dry spells when seedlings would not be able to survive. In tropical regions, where temperature and rainfall are consistently high all year round, dormancy is less important.

Dormancy tends to be a characteristic of desiccation-tolerant orthodox seeds rather than recalcitrant ones. Although some recalcitrant seeds have some dormancy, recalcitrant seeds frequently start germinating on the plant and continue to germinate after the seed has been shed. This is advantageous in the tropical and semi-tropical regions in which many plants with recalcitrant seeds originate, as competition for space, light, and nutrients favors fast growth. Anyone who has tried to grow the recalcitrant seed of an avocado (*Persea americana*) will know that simply placing it in water will soon encourage roots to grow.

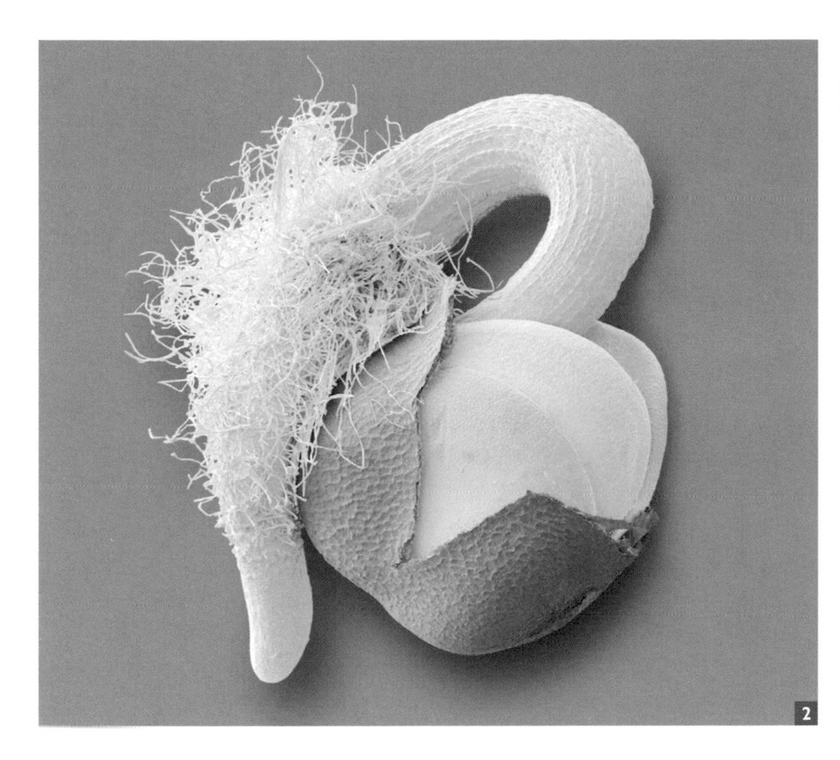

1 *A pine seedling emerges from a bed of needles on the forest floor in Spain.*

2 *This colored scanning electron micrograph captures the late stage of germination in a brassica seed. The white radicle (embryonic root) grows downwards, its hairs helping it to absorb water and nutrients from its surroundings. The orange seed coat has split to reveal the green plumule (embryonic shoot) and its seed leaves.*

TYPES OF DORMANCY

Scientists recognize different types of dormancy. Although several classification methods have been developed, one of the most widely used among seed institutions is that compiled by Carol and Jerry Baskin of the University of Kentucky, USA. Based on a system developed in 1977 by Russian scientist Marianna Nikolaeva, it classifies dormancy as physiological, morphological, morphophysiological, physical, and combinational. "The Baskins' book is considered to be the bible of seed science work," explains Rosemary Newton, Seed Ecologist for the MSBP. "It is the most comprehensive compilation of what is known about seed ecology, biogeography, and the evolution of dormancy and germination to date, and is an essential reference book for seed scientists."

THE RIGHT CONDITIONS

Physiological dormancy is the most common type of dormancy. Where present, the embryo of the seed will not grow unless certain signals are received that trigger it to release the dormancy. These are related to temperature, light, nitrates, and chemicals that regulate plant growth. The latter include abscisic acid (ABA), found in leaf litter. A prolonged period of cold or warm weather is often a requirement for germination in seeds that have physiological dormancy.

For example, scientists at the Royal Tasmanian Botanical Gardens found that seeds of *Juncus antarcticus*, gathered from an altitude of 4,300 feet (1,310 m) on Mt. Rufus in Lake St. Clair National Park, Tasmania, only germinated when exposed to alternating temperatures of 81°F (27°C) during the day and 59°F (15°C) at night after being stratified for eight weeks at 41°F (5°C). (Stratification is the term botanists use to describe treatments of seeds used to mimic natural processes to break dormancy.) The findings suggest that, in nature, the seeds' physiological dormancy breaks down during the cold and wet winter months, enabling germination in late spring or early summer once temperatures rise.

THE SHAPE OF THINGS TO COME

Morphological dormancy refers to embryos that are small (underdeveloped) and either undifferentiated (where cotyledons and radicle cannot be distinguished) or differentiated (where they can be separately identified). With this kind of dormancy, the seeds simply need time to grow.

Seeds are considered to have straightforward morphological dormancy if they germinate within 30 days. Those that take longer than this time for the radicle to emerge are considered also to have a physiological component to their dormancy and are therefore regarded as having "morphophysiological" dormancy.

Seeds from the common ash (*Fraxinus excelsior*), one of Europe's largest native deciduous trees, have morphophysiological dormancy. In the year that the tree flowers, a complete embryo develops, filling around half the seed. When the fruits are shed, the seeds become hydrated by moisture from the soil or leaf litter and the embryo grows further to fill the seed case completely. Rather than germinating immediately, however, the seed requires a period of cold temperature to reduce dormancy. Generally, germination does not occur until the second spring after the seed has formed.

BREAK THROUGH THE BARRIER

Physical dormancy is used to describe seeds that have an impermeable seed coat, and which cannot germinate until this barrier has been broken down to let water moisten the seed. In nature, physical dormancy is overcome by the seed experiencing alternating hot and cold temperatures; passing through acid present in an animal's gut; or experiencing repeated wetting and drying. This type of dormancy is exhibited by all members of the Malvaceae (mallow), Convolvulaceae (bindweed), Cistaceae (rock rose), and Rhamnaceae (buckthorn) plant families, as well as in some members of other families, such as Fabaceae (pea and bean).

The final category of dormancy is "combinational," where the seed coat is impermeable and the enclosed embryo also has some form of physiological dormancy. Some winter annuals, including some *Geranium* species, are combinational. They often relinquish their physiological dormancy in the field within a few weeks of reaching maturity but retain their physical dormancy for longer. Seeds from genera such as *Cercis* and *Ceanothus* require a few weeks of exposure to cold temperatures after their seed coat is penetrated in order to germinate.

Seeds also often respond to triggers that promote germination once dormancy has been released. For example, if very small seeds are buried very deep in the soil, they might not have sufficient reserves in their endosperm to reach the surface. Seeds can sense whether they are at a suitable depth for germination by using phytochrome, a pigment that is receptive to light. They can also detect alternating day and night temperatures if they are sufficiently close to the surface. In ecosystems where wildfires occur regularly, the presence of smoke can act as a trigger for germination; this signals to the seed that the vegetation above has been cleared and competition removed.

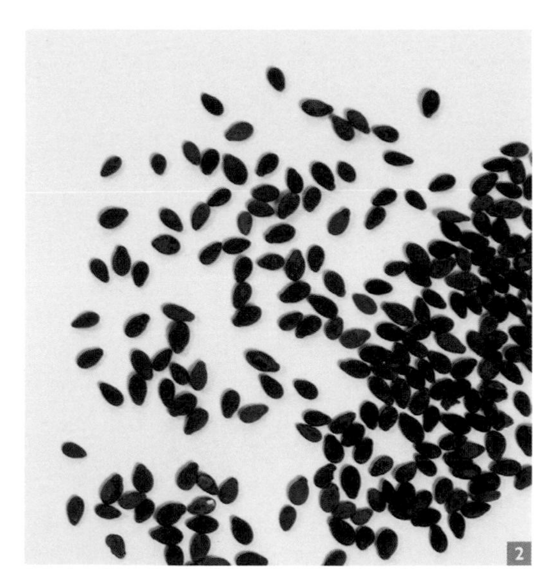

1 *Ash seedlings, seen here, only emerge after a cold spell.*

2 *These melon seeds were buried alongside Egyptian pharaoh Tutankhamun 3,300 years ago.*

TUTANKHAMUN EQUIPPED WITH SEEDS FOR THE AFTERLIFE

In the Economic Botany Collection at Kew Gardens there are glass jars containing fruits, nuts, and seeds that were buried with Tutankhamun in his Egyptian tomb some 3,300 years ago. Among the plants represented are watermelons, vines, dates, fenugreek, wheat, juniper, and coriander, which, aside from being darkened by age, closely resemble their modern-day counterparts. This botanical treasure was sent to the Gardens in 1922 by Howard Carter, the archaeologist who rediscovered Tutankhamun's tomb, then largely forgotten.

More than sixty years later, in 1988, Christian Tutundjian de Vartavan, a French student investigating ancient Egyptian plants at University College London, came across the hoard among Kew's vast collection of economic botany specimens. He studied 38 botanical samples from the tomb, located both at Kew and in the Egyptian Museum in Cairo, 29 of which were found to contain seeds and fruit. He recorded 145 species in all, of which 61 were identified to genus or species level. "The extremely dry conditions of the tomb contributed to their exceptional preservation," explains Mark Nesbitt, Curator of Kew Gardens' Economic Botany Collection.

Ancient Egyptians considered it essential to provide the *ka*, the soul of the deceased, with food similar to that consumed during their lifetime. This is why ancient Egyptian tombs often contain large quantities of foodstuffs. "The most striking aspect of the assemblage of plants is how many non-Egyptian plants there are," says Mark. "For example, almonds and juniper berries must have been imported from either the Levant or Turkey. They show us that a pharaoh would have had access to a wide range of products."

BANKING ON LIVE SEEDS

For seed banks to fulfill their objective of keeping stocks of seeds for future use, they should be placing only "viable" seeds in storage, that is, seeds that are capable of germinating. Therefore, before banking new seeds, they must test the viability of a selection of seeds from each collection. This work begins in the field with the seed collectors. Some plant families, such as Poaceae (grasses), Asteraceae (daisies), Cyperaceae (sedges), and Combretaceae, often shed high numbers of "empty" seeds, while others, such as legume seeds, can sometimes be damaged by insects. For this reason, seed collectors slice open a proportion of seeds in the field to assess their quality.

The collectors from the MSBP generally only make a collection if they consider that fewer than 30 percent of the seeds are empty or damaged, or else they collect more seeds to compensate for the non-viable ones. Once the seeds arrive at the MSBP and have undergone a period of drying to slow down the rate at which they decay, they are cleaned to remove extraneous material, such as twigs, leaves, and flowers, and, if need be, are then extracted from their fruits. Sieves and aspirators (machines that separate out light, empty seeds from heavier, full ones) help to whittle down the collection to seeds that are likely to be capable of growing.

Following cleaning, the seeds are X-rayed to see what is inside them. Healthy seeds show up as bright and dense on a gray-scale X-ray image, sometimes with the embryo clearly visible within the endosperm. Infested seeds, on the other hand, often reveal the faint outline of a beetle or other insect within the shape of the seed. The scientists involved note the proportion of viable, empty, part-empty, or infected seeds shown up by X-ray in each collection.

The real test of the seeds, however, comes when scientists attempt to germinate them. All seeds undergo germination tests when they first arrive at the MSBP and again at intervals, on average following every ten years of storage in the seed bank. Other seed banks have similar protocols. Rosemary Newton explains some of the tricks used by scientists to coax seeds into life. "We start by looking at the climate in the area the plant comes from," she explains. "We enter the coordinates of the locality from where the seeds were collected into a tool called WorldClim. It interpolates data from the weather station closest to the site to give us monthly maximum, minimum, and mean temperatures, plus rainfall statistics.

"Seeds are collected in the month they are dispersed, so we can look at whether they were shed during a warm, cold, wet, or dry period, and what the subsequent weather was likely to be. We use this information to

1

1 *A photograph (left) and X-ray (right) of seeds from the Australian hickory wattle* (Acacia penninervis). *The dark patches revealed by the X-ray show where seeds have been damaged internally, for example by nibbling insects.*

2 *A seed-germination expert uses a thermo-gradient plate to test seeds' ability to germinate under different temperature regimes.*

work out the conditions that might trigger germination and whether the plant would be likely to have some form of dormancy. We also look at the taxonomy, as there are sometimes trends within families; for example, with a bean species, we know we will probably need to overcome its physical dormancy. We also search the academic literature to see if any work has already been carried out on the plant. And we look at the MSBP's Seed Information Database (SID) which contains information on any previously conducted germination tests."

TRICKS OF THE TRADE

Armed with this information, the scientists play nature at its own game by placing seeds in conditions of temperature and light that they think will induce germination, with scarification by a scalpel or sandpaper being used to penetrate the coat of seeds with physical dormancy. The seeds are then placed in a Petri dish on a thin layer of agar jelly within one of several walk-in incubators within the MSBP's state-of-the-art laboratories. Inside, they are subjected to a fixed temperature ranging between 41°F and 104°F (5°C and 40°C) but most commonly 50°F or 59°F (10°C or 15°C). Sometimes, large seeds are immersed completely in moist sand, rather than being placed on agar, so there is no chance of them drying out during the experiment.

The seeds are left for at least 42 days, often longer. If there's no sign of germination after this time, the scientists might try moving the seeds to a neighboring incubator set at a higher or lower temperature, or they might expose the seeds to alternating hot and cold temperatures, representing the warming and cooling of daytime and nighttime. If results are needed quickly, the scientists can use hormones such as gibberellic acid to trick the seeds into thinking they have experienced, say, a cold period. In plants, gibberellic acid promotes growth and the elongation of cells, enabling them to grow towards sunlight.

REFUSING TO GROW UP

Some seeds have proved particularly reluctant to germinate away from their natural environment. Rosemary had a taxing time while studying for her Master's degree in 1999 trying to grow plants of *Cannomois virgata*, commonly known in its native South Africa as *besemriet* (Afrikaans for "broom reed"), as the stems are often used for making brooms. This plant is a member of the Restionaceae family, which, together with species of Proteaceae and Ericaceae, make up a large proportion of South Africa's Cape fynbos habitat. The fynbos is a type of heathland that grows across the mountains, valleys, and coastal areas of the country's southern and southwestern Cape. It is the major vegetation type of the Cape Floral Region, one of the world's 18 biodiversity "hotspots." This region represents less than 0.5 percent of the area of Africa but is home to nearly 20 percent of its flora.

Rosemary suspected that dormancy, rather than inappropriate collection and storage methods, lay behind the repeated failed attempts to germinate *C. virgata* recorded in scientific literature. The fynbos ecosystem is characterized by Mediterranean-like rainfall in winter and drought in summer, with periodic wildfires. The soils are generally poor in nutrients. Any attempts to germinate a plant from this ecosystem would therefore need to emulate these seasonal rhythms. "I conducted all kinds of experiments to try and get the *C. virgata* seeds to germinate, but they were not successful," recalls Rosemary. "Then the very last one, at the end of my two year Masters, finally resulted in around two thirds of the seed germinating. I was very pleased."

1

Key to Rosemary's success was being able to emulate a changing sequence of environmental conditions culminating in a fire event. Wildfires are a natural part of the fynbos ecosystem but they only happen every seven to twelve years, so seeds must be able to survive in the soil for that long to germinate successfully. It has been observed that nut-fruited Restionaceae, of which *C. virgata* is one, regenerate after fire events, both by seeds germinating and by re-sprouting. However, if seeds stay on the soil surface for any length of time they get eaten by rodents. The Restionaceae seeds have avoided this by developing a symbiotic relationship with ants. The ants pick up the seeds and carry them to their nests where they eat off the seed's elaiosome. The seeds lie 1½–2¾ inches (40–70 mm) below the surface, out of reach of predators and deep enough to survive the intense heat of a fire that would kill them at the surface.

RESPONDING TO TREATMENT

Rosemary experimented on two sets of *C. virgata* seeds: one stored in brown paper bags in ambient, dry laboratory conditions, and the other buried in seed trays ½ inch (13 mm) deep in fynbos soil gathered from Silvermine Nature Reserve in Cape Town. She divided these up so that one third of the seeds in each set was used as a control against which to measure results in the others; another third was treated with fire; and the final third was exposed to charate (powdered charred wood). The seeds buried in soil underwent a regime of various alternating warm and cold temperatures and wetting over a 16-week period, emulating the day and night temperatures and light exposure that would ordinarily be experienced by the seeds in the wild. The "fire" treated seeds were buried in soil, covered with fynbos brush and then ignited, thus exposing the seeds to heat. Once the soil had cooled, a mixture of soil and surface charate was placed into trays, into which either fire- or charate-treated seeds were sown.

The highest rate of germination was achieved in those seeds stored in soil that were then exposed to either charate or fire, following a prolonged period of alternating warm and cool temperatures and periodic wetting. In both these samples, around 65 percent of seeds germinated, suggesting that the heat from fire had little impact on germination but that chemicals released by the fire provided signals for the seed. Germination from all the soil-stored seeds was considerably higher than that from the seeds that had been stored in the laboratory. The findings suggest that seed dormancy in *C. virgata* breaks down slowly during burial in the soil, and that germination only takes place once a fire burns down the vegetation above. Many seeds can deteriorate if subjected to repeated drying and wetting, but the Restionaceae most likely have built-in repair mechanisms that help seeds remain viable during long periods lying dormant in the soil.

1 *South Africa's Cape fynbos habitat, a type of heathland where* Cannomois virgata *grows.*

2 *In the fire stage of the experiment, seeds were buried in soil. Sand fynbos vegetation was placed on top and then lit.*

THE TEST OF TIME

In 1879, the American botanist William J. Beal, founder of the W. J. Beal Botanical Garden at Michigan State University, began an experiment to see how long seeds of some common local plant species might remain dormant in the soil without losing their ability to germinate. He planted fresh seeds in uncorked pint bottles containing moist sand, and buried them in the grounds of the university. Each bottle contained 50 seeds from the following species: *Agrostemma githago, Amaranthus retroflexus, Ambrosia artemisiifolia, Anthemis cotula, Brassica nigra, Bromus secalinus, Capsella bursa-pastoris, Erechtites hieracifolia, Euphorbia maculata, Lepidum virginicum, Malva rotundifolia, Oenothera biennis, Plantago major, Polygonum hydropiper, Portulaca oleracea, Rumex crispus, Setaria glauca, Stellaria media, Thuja occidentalis, Trifolium repens, Verbascum blattaria*, and *Verbascum thapsus*. The bottles were placed neck down to avoid waterlogging. Over the following 40 years, one bottle was opened every five years and germination tests conducted on the seeds. When Beal's successor, Henry Darlington, took over running the experiment in 1915, he increased the period between tests to ten years so as to extend the experiment, an interval extended again, this time to 20 years, by the experiment's next overseers, Aleksander Kivilaan and Robert Bandurski. The 15th bottle, opened in 2000, the experiment's 120th year, yielded the seedlings of two species, *Verbascum blattaria* and *Malva neglecta* (syn. *Malva rotundifolia*). *Verbascum* is the only plant to have germinated in every year of the trial.

"Professor Beal often attended farmer institutes in the state of Michigan back in the late nineteenth century," explains Frank Telewski, Professor of Botany at Michigan State University, Curator of the W. J. Beal Botanical

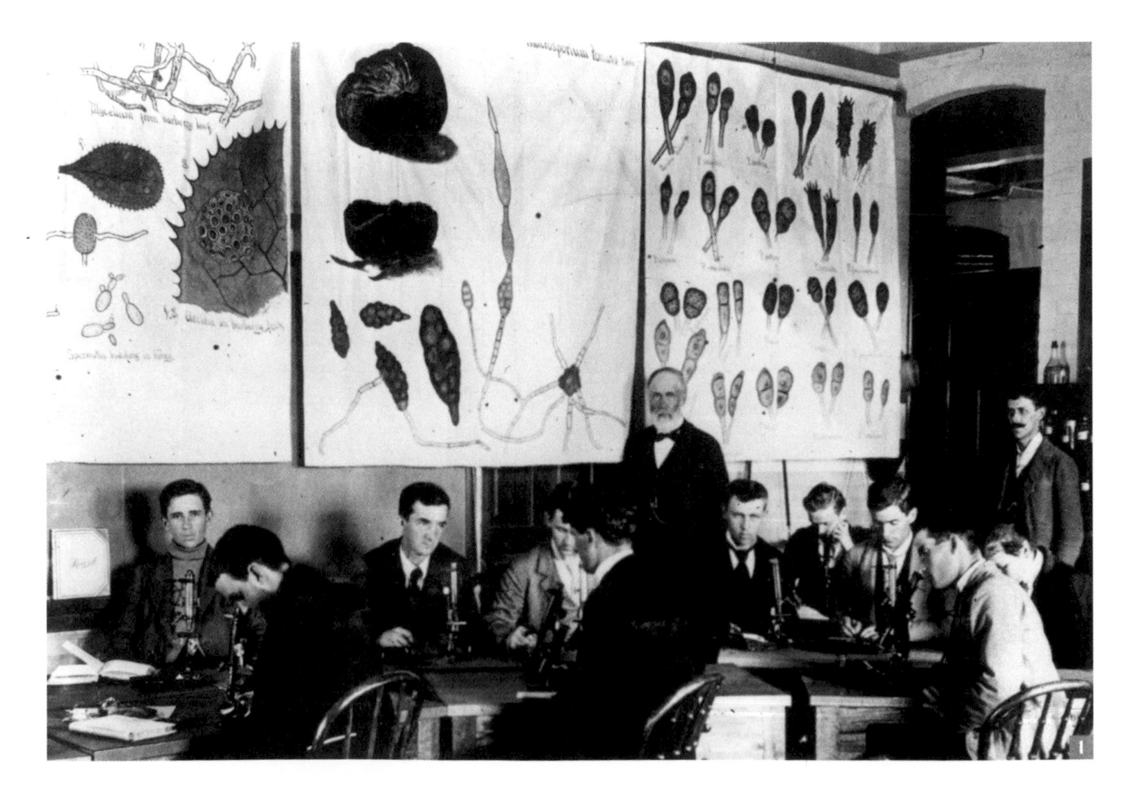

1 *William J. Beal (standing, center), who began an ongoing study into seed germination in 1879 at Michigan State University, USA.*

2 *Beal buried 20 sand-filled glass bottles, each containing 50 seeds of common local plants. Researchers open a bottle periodically to see which seeds are still capable of germinating.*

Garden, and present overseer of the experiment. "In those days, farmers didn't have chemical herbicides or genetic engineering; they had to remove weeds manually with a hoe. It was hard work, so they wanted to know how long the seeds would remain viable in the soil." Their curiosity led Beal, who was Professor of Botany at the time, to begin his experiment. He buried 20 bottles at a depth that was great enough for the seeds not to be exposed to enough light to make them germinate but would still allow exchanges of moisture, oxygen, and carbon dioxide between seeds and soil.

DUG OUT OF DORMANCY

With weed seeds, it is often light that triggers germination, causing weeds to appear after ploughing. The common poppy (*Papaver rhoeas*) is a case in point. Able to lie dormant for 80 years before germinating, poppy seeds respond to light when soil is disturbed. During the First World War, hundreds of poppies sprang to life on battlefields that had been churned up by fighting soldiers, giving rise to the flower's enduring status as a symbol of remembrance.

HOPE FOR LOST HABITATS

"Beal's experiment, which is set to run for another century, has shown that seeds can remain viable in the soil for upwards of 120 years," says Telewski. "With the moth mullein, *Verbascum blattaria*, it not only remains viable in the soil but persists with a really high viability; when we opened the last bottle in 2000, we had a 50 percent germination rate for *V. blattaria*."

Local farmers, still fighting weeds, have reported similar seed longevity. Southwest Michigan used to be covered in tallgrass prairie but when the first European settlers arrived in the late seventeenth century, they began to turn it into agricultural land. "It was easier for them to plough grasslands than raze forests," says Telewski. Now, less than 1 percent of natural tallgrass prairie habitat remains in Michigan. However, when one farmer excavated a pool and piled up the spoil on its banks, the old tallgrass prairie began to grow. "Species that hadn't been seen on his farm for 120 or 130 years were still viable," says Telewski. "Think of the implications that has for restoring habitats. If you are restoring a habitat and

1 *This scanning electron micrograph reveals tessellating hexagons on seeds of the common poppy* (Papaver rhoeas).

2 *Common poppy seeds germinate when they are exposed to light as ground is disturbed. Poppies grew profusely after soldiers churned up soil when fighting during World War I.*

3 *The moth mullein* (Verbascum blatteria) *was among seeds buried in 1879 by William J. Beal. When one of Beal's bottles of seeds was opened in 2000, half of the seeds of* V. blatteria *were still able to germinate.*

the nearest extant population is 250 miles (400 km) away, the genetic makeup of that population may be different to the one you are trying to restore. If you carry out DNA analyses on seeds from the local soil bank and living populations, you can see how much variability exists between those two natural populations, and ensure the restored habitat is genetically appropriate."

Such knowledge could be very valuable when it comes to restoring Michigan's tallgrass prairies, as efforts so far have failed to replicate remnant, untilled prairie habitats. Work to restore prairies usually involves sowing seeds of native species and encouraging wildfires on former agricultural land. Such activities have resulted in plant communities with abundant native species that support prairie insects and animals but in which the species mix differs from the original prairie ecosystem. Being able to tap into seed banks within the soil of the land to be restored could help resolve this difference. Further research in this field could help scientists better understand the complex ecological interactions that take place among plant and animal assemblages, as land recovers from agricultural and other practices.

PRESERVED FOR POSTERITY

Allegedly, the oldest tissue ever to have yielded a living plant was 32,000 years old. In 2012, Russian scientists claimed to have grown the seedling of *Silene stenophylla*, a white-flowered plant native to Siberia, after discovering seeds preserved in ice 130 feet (40 m) beneath the permafrost. Unearthed among layers containing bones from mammoth, bison, and woolly rhinoceros, they had been buried in the burrow of an ice-age ground squirrel. Mature seeds from the hoard had been damaged, but immature ones contained sufficient viable tissue for cultivation. Plants grown from the tissue had white flowers, identical to each other but different from modern plants of the species.

This remarkable feat of plant husbandry, although not independently verified, indicates that it might be possible for seeds and other tissue to produce living plants even when they have been frozen for an extremely long time. It may even be possible to revive plants that have become extinct. The oldest intact seed to germinate was a 2,000-year-old date palm seed from Masada in Israel. Nicknamed Methuselah, the plant appears to be living up to its name. Since being planted in 2005, it has grown to more than 8 feet (2.5 m) tall and flowered several times.

SEIZED SEEDS SPROUT AFTER TWO CENTURIES

In 2005, the Dutch researcher Roelof van Gelder was exploring the Dutch High Court of Admiralty prize papers held at the National Archives in Kew, UK, when he came across a red leather wallet containing packets of seeds. The notebook was embossed in gold with the name of Jan Teerlink, a merchant from Vlissingen (known as Flushing in English), Holland. The paper envelopes, meanwhile, were mostly labeled with the Latin names of the seeds they contained.

The notebook was found among documents that had been on board the ship *Henriette* when it was seized by the British in 1803, when Holland was an annex of Napoleon's France. Britain and France were at war, so the cargo of a Dutch trader was legitimate plunder.

Documents revealed that *Henriette* had traveled to China, where tea, porcelain, and silks had been loaded on board. On its homeward journey to Holland, the ship had anchored off Cape Town, South Africa, where the Dutch had a colony. Teerlink spent two weeks touring the sights. He probably acquired the seeds from The Company's Garden, as they were wrapped and labeled in a manner that was typical of botanists. This botanical garden, which still exists, was established in 1652 by Teerlink's former employer, the Dutch East India Company.

Staff at the National Archives wondered if the seeds could be grown into plants, so they contacted Kew Gardens, located just down the road. The 40 packets of seeds turned out to contain 32 species, including heathers, daisies, watermelons, legumes, and proteas. They all came from the fynbos vegetation that is unique to the Cape. Matt Daws, seed ecologist at the MSBP at the time, was given the task of trying to germinate them.

Although the seeds had been kept in relatively cool conditions at the National Archives, the level of humidity they were exposed to was quite high. There were not enough seeds in the collection to experiment with finding out which conditions would most likely prompt germination, so it was a case of taking a best guess. This involved exposing the seeds to alternating temperatures of 68°F and 50°F (20°C and 10°C), so as to emulate the diurnal warming and cooling they would experience in autumn and early winter, when most plants in the fynbos germinate.

Because wildfires are natural events in the fynbos ecosystem, Matt also exposed them to smoke. This was done by soaking seeds in "smoke water," made by a process that involves bubbling smoke from burning fynbos vegetation through water. The seeds were then sown onto agar jelly. Meanwhile, he chipped the hard coats of the legume seeds to overcome their physical dormancy and let water in.

In March 2006, 16 seeds of *Liparia villosa* germinated, followed by one each of an *Acacia* and a *Leucospermum*. The *Leucospermum* survives to this day, a healthy 3 feet- (91 cm-) high bush that is testimony to the ability of seeds to survive for long periods, even when stored in less-than-ideal conditions.

1 *Senior research associate Svetlana Yashina inspects an outgrowth of* Silene stenophylla, *possibly the oldest plant ever to be regenerated, at a laboratory of the Institute of Cell Biophysics, Russian Academy of Sciences, in Pushchino, Russia. The plant was grown from tissue found in an ice-age ground squirrel's burrow.*

2 *Jan Teerlink's leather notebook and the packages of seeds it contained.*

INSIDE A SEED

Botanists have traditionally distinguished two types of flowering plants. Emerging DNA techniques have enhanced this classification but it remains a good way to distinguish different types of seeds.

INSIDE THE SEEDS OF FLOWERING PLANTS

To understand exactly what seeds comprise, it is helpful to know how flowering plants are classified. The angiosperms have traditionally been divided into two classes, based on fundamental differences in their form or life cycle. The divisions are Magnoliopsida or dicotyledons, 'dicots', and the Liliopsida or monocotyledons, 'monocots'. The seedlings of dicots have two seed leaves, or cotyledons, in the embryo of the seed. Those of monocots have only one seed leaf. There are about 175,000 species of dicots. They include most common garden plants, shrubs and trees, as well as broad-leafed flowering plants, such as magnolias, roses, geraniums, and hollyhocks. There are around 75,000 species of monocots, encompassing orchids, daffodils, lilies, irises, palms, grasses, and sedges.

BEAN SEED (DICOT)

SEED COAT

Formed from a fertilized ovule, the seed coat, or testa, encloses and protects the embryo of the future plant.

COTYLEDON

Cotyledons are destined to become the first leaves of the seedling. Once a seed has grown to its full size, developmental changes continue at cellular level. Embryonic cells within cotyledons begin to store proteins, lipids and starch; these provide energy and basic building blocks needed, following a period of dormancy, for the seed to germinate and for a seedling to grow. The cotyledons of dicots are able to photosynthesize so they are similar to true leaves. The cotyledons of monocots comprise a scutellum and coleoptile (ligular sheath). The scutellum absorbs food from the endosperm, while the coleoptile covers the plumule that grows to become the stem and leaves of the plant.

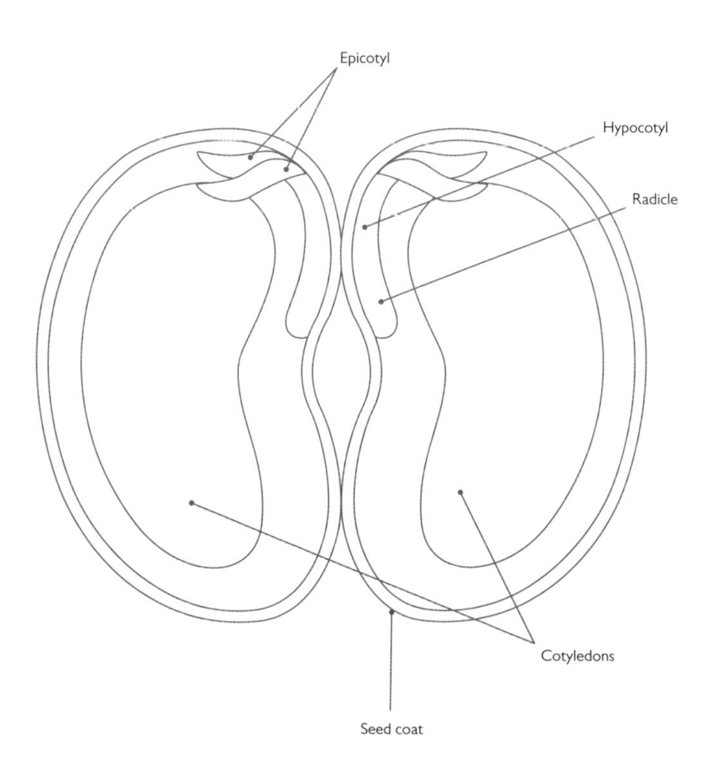

DICOT AND MONOCOT SEEDS

It is relatively easy to distinguish the seeds of dicots from those of monocots. Dicot seeds, such as those from runner bean plants, split into two parts. Enclosed within a soft outer coat that comes off quite easily, these two cotyledons provide food for the embryo.

Monocot seeds, such as those from maize plants, have only one part. They, too, have an outer coat but it is not shed so easily. A single cotyledon, the endosperm, surrounds the embryo and provides food for the seedling.

SHIFTING CLASSIFICATIONS

The emergence of DNA testing has provided new methods by which to classify plants. According to APGIII, the latest classification of flowering plants published by the Angiosperm Phylogeny Group, monocots form a clade (a group comprising an ancestor and all its descendants). APGIII recognizes another clade called eudicots, which contains the majority of dicots. The remaining dicots are sometimes called palaeodicots; they are differentiated from the eudicots by their pollen structure. Some herbariums are now using this new classification to arrange their dried plant specimens. New relationships are emerging from the application of sequencing techniques. For example, the corpse flower (*Rafflesia*), which grows in Indonesia and has the largest individual flower of all plants, is related to the poinsettia (*Euphorbia pulcherrima*), which has one of the world's smallest flowers.

MAIZE SEED (MONOCOT)

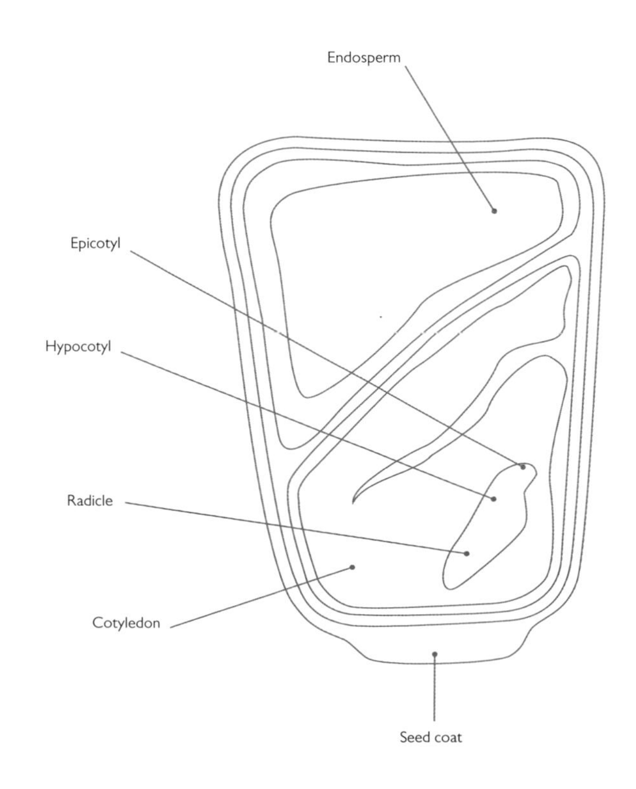

ENDOSPERM

The endosperm forms during the process of double fertilization. One sperm fuses with the egg to form a zygote, while the other fuses with the two polar nuclei of the central cell, to form a triple fusion; the endosperm. This tissue, found in monocots and many dicots, acts as the larder for the future plant, storing starch, oils and proteins that are broken down during germination. It comprises three cell types: the basal transfer layer that carries nutrients from the mother plant as the seed develops; the starchy endosperm, where the starch, oil and proteins are stored; and the aleurone, which secretes amylase and helps break down the starchy endosperm to provide sugars that act as fuel for the seedling.

HYPOCOTYL AND EPICOTYL

The hypocotyl is the stem that connects the radicle to the cotyledons; the epicotyl is the stem above the cotyledons and below the first true leaves.

RADICLE

The radicle, located at the lower end of the embryo, forms the root of the seedling. Dicot radicles produce an apical meristem, which generates root tissue during the plant's life. Monocot radicles, by comparison, are aborted, and new roots sprout from nodes in the stem.

S E E D B A N K U K

INSPIRING SEED BANKING AROUND THE WORLD

NAME

Kew Gardens' Millennium Seed Bank Partnership

NUMBER OF ACCESSIONS

Seeds of some 36,000 wild plant species are held in the MSBP's vault. As of August 2015, the collection comprised just over 2 billion individual seeds.

WHEN FOUNDED

The MSBP opened in 2000, although seed science began at Kew Gardens as far back as 1898. The Gardens' first dedicated seed-collecting expedition was to the Mediterranean in 1974.

FOCUS OF THE COLLECTION

The MSBP focuses on collecting seeds from wild plants that are: endemic, endangered (by threats such as land-use change, logging, and climate shifts), or economically valuable. It aims to bank seeds from 25 percent of all wild plant species by 2020. Plants from dryland, alpine, coastal, and island habitats are a particular focus, as these are the most vulnerable to climate change.

WHY IT IS NEEDED

Between 60,000 and 100,000 species of plant are faced with the threat of extinction. Among these are plants that are valuable as food, medicine, building materials, and clothing, as well as new species that have yet to be found.

WHO FUNDS IT

Two key sponsors helped bring the MSBP to fruition: the Wellcome Trust and the Millennium Commission. Today, it receives funding from the UK Government and a variety of private sponsors.

1 *Kew Gardens' MSBP, which has so far banked seeds of 36,000 wild plant species.*

2 *Displays outside the MSBP educate visitors about different habitats in the UK.*

3 *Large seeds stored in glass jars at the MSBP are reminiscent of candies in a candy store.*

WHERE SEEDS ARE STORED

The seeds reside in three walk-in freezers within a subterranean vault beneath the $27.3 million (£17.8 million) Wellcome Trust Millennium Building at Wakehurst Place in West Sussex, UK, the design of which was inspired by the undulating seedpod of a sea bean. Protected by a heavy safe door and airlock, the vault is designed to last 500 years. Emergency pumps are at the ready should groundwater levels rise, while a sensor on the roof records radiation levels.

In the airy visitor center, one of four tunnel-shaped sections of the main building, more than 60 staff catalog, clean, document, germinate, and store for posterity seeds that arrive from all over the world.

Visitors can observe the seeds' progress around the building through windows into the laboratories. After arriving in the staging area of the Herbarium room, specimens are placed in an initial drying room. From here they pass to the cleaning room, where detritus is removed; then it's back to the drying room, where the cool circulating air slowly desiccates the specimens as they sit in cotton or paper bags. This process completed, they are X-rayed to check their quality, and then counted into lots of 50 seeds and weighed to ascertain the average seed weight. Each accession is split into two or three collections and divided between the freezers. Only seeds stored in the "Active" freezer are accessed for testing or use; the others are left untouched.

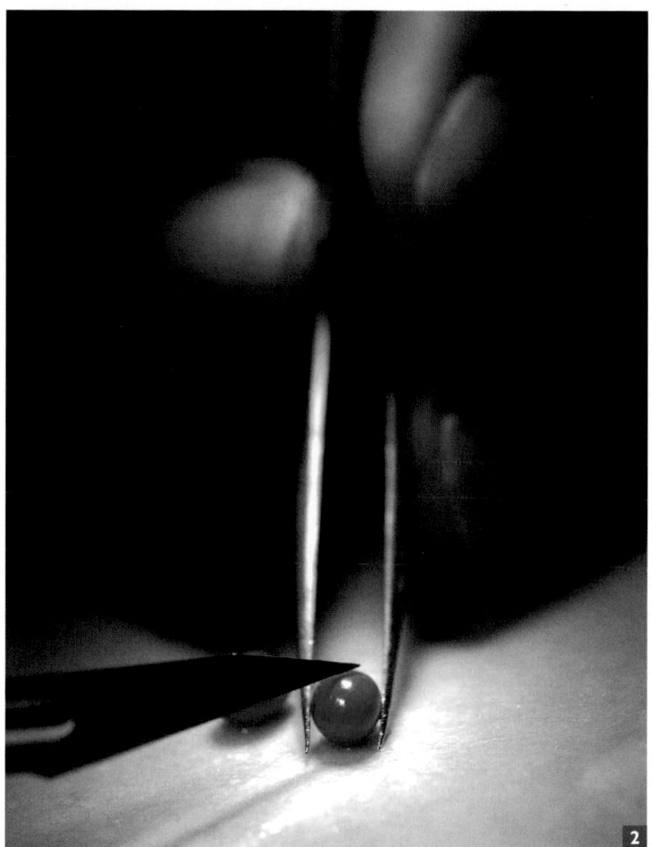

1 *Seeds arriving at the MSBP are initially stored in cotton bags in a drying room.*

2 *Scarifying seeds, by cutting into their seed coats with a scalpel, helps to overcome physical dormancy and facilitate germination.*

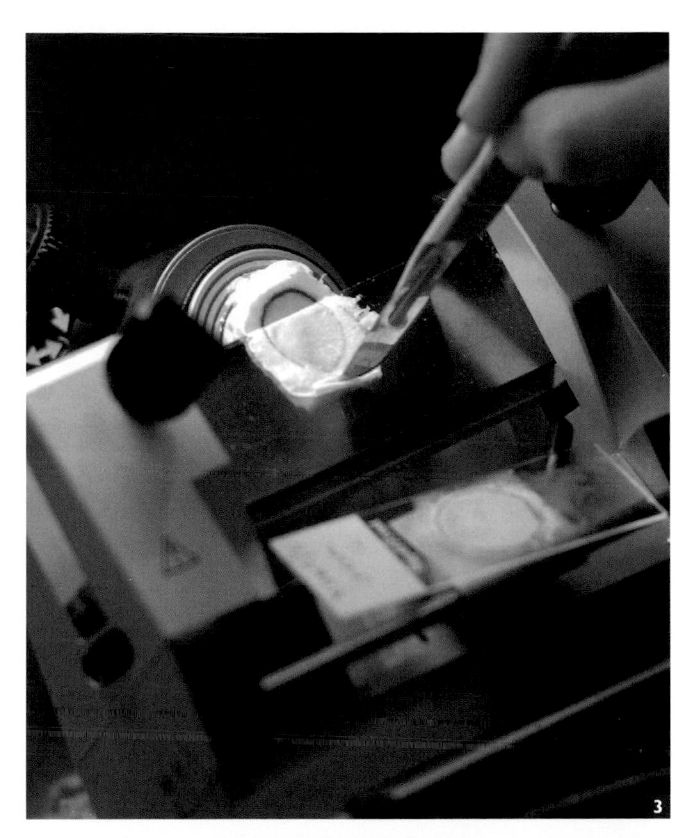

3 *Researchers make thin sections of seeds so they can be viewed under a microscope.*

4 *Microscopes at the MSBP are linked to computers, enabling scientists to examine seed structures in detail on screen.*

CURRENT RESEARCH

The MSBP conducts a wide-ranging research program that investigates how seeds respond to their environments, how they shift from dormancy to germinating, and how seed chemistry varies from one specimen or species to another. The organization works in close collaboration with partners in 80 countries. Major projects range from the Useful Plants Project, which helps communities in Botswana, Kenya, Mali, Mexico, and South Africa to grow and sustainably use indigenous plants, to the UK National Tree Seed Project, which is gathering seeds from all UK tree species so woodlands can be better managed and conserved.

SEEDS WITH A STORY

In 2009, the Yunnan banana *(Musa itinerans)* became the 24,200th species to have its seeds banked by the MSBP. In gathering the seeds of this pink banana, an important food for Asian elephants, the MSBP attained its goal of saving 10 percent of the world's flora by 2010.

PLANTS AND SEEDS FROM THE WORLD'S COASTAL ZONES

Coastal habitats, such as coral reefs, seagrass meadows, salt marshes, and mangrove forests support marine biodiversity, protect shores against flooding and act as nurseries for fish and other marine organisms. They are among the most valuable ecosystems on Earth but are disappearing faster than global rates of rainforest loss. Here are some particularly vulnerable and useful species.

CANAVALIA ROSEA

Canavalia rosea is a perennial climber, with flattened sausage-shaped seedpods and rose-colored flowers, which dwells on seashores across the Tropics. It forms mats on beaches close to the high-tide mark, which help prevent coastal erosion. It has recently been investigated as a potential foodstuff, particularly in developing countries where malnutrition prevails. The brown, streaky seeds of *Canavalia rosea* are rich in proteins, carbohydrates, dietary fiber, and essential amino acids; however, they are poisonous if eaten raw.

This is perhaps why the philanthropist and naturalist Joseph Banks described them as "a kind of beans, very bad," after eating them in 1770 while plant-collecting during Captain Cook's circumnavigation of the globe. *C. rosea*'s fast growth, large volume of seeds, tolerance to salinity, and resistance to pests, in fact, make the plant potentially valuable within agriculture. Researchers at Mangalore University in India suggested in 2014 that varieties of *C. rosea* could be used to boost nutritional health in both humans and livestock.

SCAEVOLA PLUMIERI

Scaevola plumieri, the inkberry, is a pioneer plant (a species that is the first to colonize habitats, paving the way for other plants also to become established). Tolerant of drought, wind, and salt, it lives on many tropical and subtropical sand dunes in Africa, the Mascarenes, Sri Lanka, Florida, and tropical South America. It can be largely subterranean, with only the

1 Canavalia rosea *forms mats on beaches, close to the high-tide mark. Its roots help to bind the sand together, preventing erosion.*

2 *The survival of the stinking hawksbeard* (Crepis foetida) *is uncertain in the UK.*

tips of its branches visible above the sand's surface; this makes it good at stabilizing sand dunes. The United States Department of Agriculture recommends its use for coastal dune restoration. However, in some areas, including Florida, *Scaevola plumieri* is endangered, threatened in particular by the invasive species *Scaevola sericea*, a vigorous coastal shrub from the Indo-Pacific, which was introduced to Florida as a landscaping plant.

POLYGONUM MARITIMUM

This member of the knotgrass family (Polygonaceae) is endangered in the UK, where it is found at the northern extent of its Mediterranean range. It grows just above the high-tide level and thus is vulnerable to higher-than-average tides. However, in recent years it has become more widespread in England, possibly encouraged by a series of mild winters and warm summers. Scientists believe that it may have potential as an indicator species for climate change. *P. maritimum* appeared on a set of first-class stamps highlighting endangered plant species in the UK, issued by Royal Mail in 2009.

SALSOLA NOLLOTHENSIS

This succulent species, also known as the salt bush, grows in hummocks on the sand dunes of the Sperrgebiet region of southwest Namibia. In the 1980s, some populations of the plant were destroyed when trenches 130 feet wide (40 m) were bulldozed for diamond mining. The plants' subsequent failure to regenerate over the following 20 years was detrimental to local ecosystems, as salt-bush hummocks help support other plant life. Its demise threatened plants such as the woody succulent species *Galenia pruinosa* and the bristly lovegrass (*Cladoraphis cyperoides*), as well as the desert rain frog (*Breviceps macrops*) and the Namaqua dwarf adder (*Bitis schneideri*).

When further mining was proposed by the company Namdeb, scientists from Kew Gardens and the MSBP investigated ways of restoring the *S. nollothensis* habitats once mining had finished. They found that the species presents a naturally high occurrence of empty seeds as well as having short seed longevity, and that deep burial of seeds prevents seedlings from emerging. This led them to conclude that reseeding was not a feasible method for restoring the habitat, and instead it was proposed that the seedlings should be transplanted.

CREPIS FOETIDA

Otherwise known as the stinking hawksbeard, this member of the daisy family (Compositae) was thought to have become extinct in England after 1980, the year of the last recorded specimen at Dungeness on the coast of Kent. However, thanks to seeds collected there in 1978, several populations were re-established in 2008. Two of these, at Rye and Northiam in East Sussex, went on to produce more than three generations of plants. The species, which has dandelion-like flowers and smells of bitter almonds, favors disturbed shingle and rocky outcrops. It still grows widely in Europe but is on the decline in western and central areas, and is now endangered in the Netherlands.

WOOD ANEMONE
(ANEMONE NEMOROSA)

GENUS Anemone
FAMILY Ranunculaceae
SEED SIZE Has flask-shaped achenes, each ¼ inch (5 mm) long
TYPE OF DISPERSAL Animal (ants)
SEED STORAGE TYPE Orthodox
COMPOSITION Data not available

For seed banking to be credible as a form of ex-situ conservation, scientists must be able to grow stored seeds into plants, possibly hundreds of years into the future. However, this will only be possible if the seeds are capable of surviving for that long. The MSBP seeks to determine the longevity of its seeds by carrying out periodic germination tests and by conducting experiments that simulate aging. One achene that has proved particularly tricky to store for any length of time is that of the wood anemone (*Anemone nemorosa*).

When MSBP scientists attempted to germinate seeds of the pretty white-flowered perennial, which had been collected in the UK just a few years earlier, they found that all of them were dead. On studying the seed in greater detail, they found that, at the time of dispersal, the embryo acts like that of a recalcitrant seed. However, following dispersal, the embryo quickly differentiates and becomes tolerant to desiccation, in the same manner as a typical orthodox seed. At this point, *A. nemorosa* seeds can be dried and stored.

"If they are left in leaf litter, however, the embryo begins to grow inside the seed, almost as if the plant is germinating within its seed," explains Robin Probert, Head of the MSBP UK Conservation and Restoration Group. "Once that starts to happen, the seeds lose their ability to withstand drying and become recalcitrant again. Even if stored within that short window, they remain incredibly short-lived. We think they will only survive in the seed bank for a few years at best. They are right at the transition between recalcitrant and orthodox."

The widely accepted Baskin and Baskin classification of dormancy would consider seeds of the wood anemone to have morphophysiological dormancy. However, Robin and colleagues at the MSBP believe that such a classification is inappropriate in the case of this species and many others that have tiny embryos at the time of dispersal. This is because, in nature, the embryo starts to grow immediately after dispersal and continues to grow during the summer months until the radicle emerges in the autumn. Because there is no arrest in development, at least in the root end of the embryo, the term dormancy seems to be inappropriate.

"Interestingly, the epicotyl, the shoot end of the embryo, *is* dormant," says Robin. "When an anemone seed sends out its radicle, the shoot remains in a state of developmental arrest until it's gone through winter chilling. That doesn't grow until the following spring. That to me is true dormancy. So I'd say that the radicle is *not* dormant but the epicotyl *is* dormant." Either way, the finding that the anemone is difficult to bank and germinate has implications for the long-term conservation of this species and, potentially, many other temperate woodland plants.

The elegant wood anemone (Anemone nemorosa), *flowering in the Devon countryside of the UK. The anemone's seeds have proved challenging to store at the MSBP, as they are only tolerant to drying for a short period of time.*

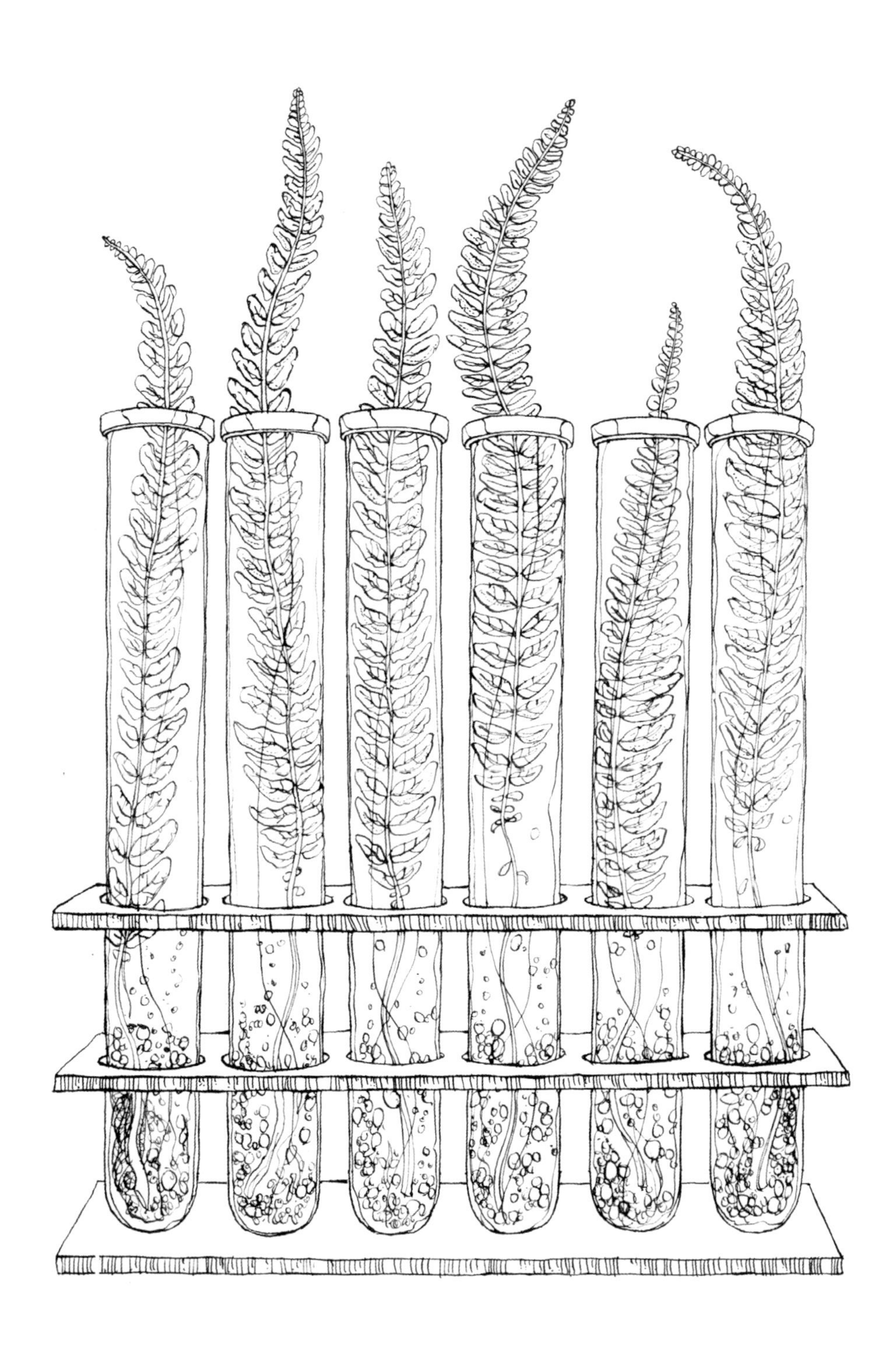

CHAPTER 6

USING SEEDS TO ENSURE HUMANITY'S SURVIVAL

SAVING CULTIVATED AND WILD SEEDS

The phenomenon of seed banking has emerged largely since the 1980s and developed rapidly around the world. Although farmers, gardeners, and horticulturalists have long collected seed for their own uses, increasing environmental concerns have driven the rise of more organized seed-storing initiatives. Today, the Food and Agriculture Organization (FAO) of the United Nations estimates that there are 1,750 individual gene banks worldwide (holding seeds, along with bulbs and tubers). Around 130 of these hold more than 10,000 accessions.

Seed banks were established in the USA and Russia early in the twentieth century. Botanists realized that selective breeding by farmers down the centuries had gradually reduced the genetic diversity contained within crop seeds, and sought to save their genetically rich wild crop relatives for use in agriculture. By the 1980s, scientists understood that wild plants were also under threat, from logging, urbanization, and climate change. This concern gave rise in 1992 to the United Nations Conference on Environment and Development, known as the Rio Earth Summit, which was attended by representatives of 172 governments.

The Convention on Biological Diversity, which was launched at the conference, suggested in Article 9 that contracting parties should: "Take action where necessary for the conservation of biological diversity through the in situ conservation of ecosystems and natural habitats, as well as primitive cultivars and their wild relatives, and the maintenance and recovery of viable populations of species in their natural surroundings, and implement ex situ measures, preferably in the source country." The call for "ex situ measures" was the catalyst for existing seed-banking initiatives to be strengthened and new operations to be set up. Today, seed banks exist both to underpin food security and to preserve natural biodiversity.

THE DROP IN THE NUMBER OF US CROP VARIETIES BETWEEN 1903 AND 1983

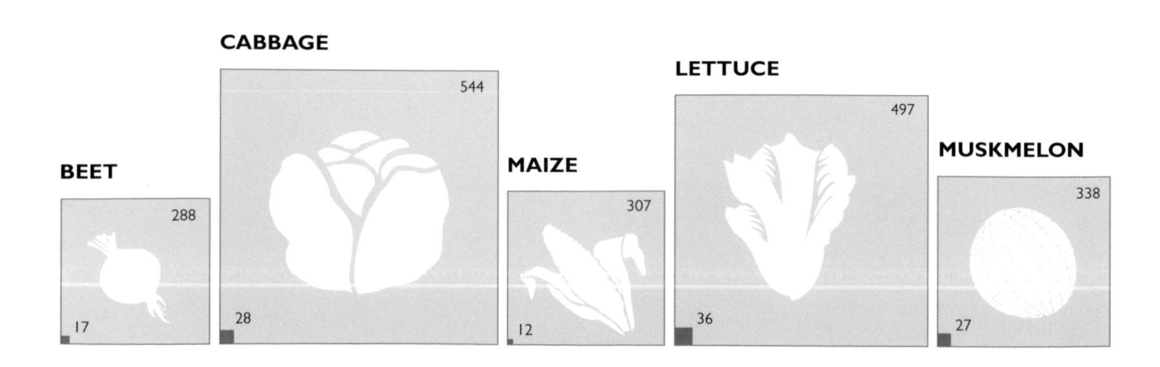

SAVING SEEDS TO CONSERVE CROP VARIETY

Agricultural seed banks focus on conserving landraces and cultivars. Among them are those held by member organizations of the global agricultural research partnership known as CGIAR (see page 174). Agricultural seed banks, whether large national centers or small community initiatives, are driven by the need to conserve varieties of crops. The FAO estimates that 75 percent of crop biodiversity has already disappeared from the world's fields. In the USA, an estimated 90 percent of historic fruit and vegetable varieties have vanished. In the Philippines, the thousands of rice varieties once grown have dwindled to fewer than 100. And in China, 90 percent of wheat varieties grown 100 years ago no longer exist.

In 2008, the Svalbard Global Seed Vault opened on Spitsbergen, a remote Arctic island. This "doomsday vault" acts as a backup for all agricultural seed collections around the world. Managed jointly by the Norwegian Government, an association of Nordic gene banks, and the Global Crop Diversity Trust, it represents the world's largest concentration of crop diversity. It has the capacity to store some 4.5 million crop varieties, each represented by a collection of around 500 seeds (a total storage capacity of 2.5 billion seeds). Currently it stores, at 0°F (-18°C), 830,000 accessions ranging from maize, rice, and wheat, to aubergine, chickpeas, and lettuce.

THREATS FROM WAR AND NATURAL DISASTERS

The need for such a backup has already been demonstrated. In 1996, a flood following a typhoon in the Philippines left its national collection of crop seeds under water; two years later, Hurricane Mitch destroyed Honduras's seed bank. But wars pose as great a threat to seeds as natural disasters. In 2002, two stores holding seeds from Afghanistan's native crops were destroyed when thieves emptied the plastic containers in which they were stored, rendering the collection useless. Conflicts have also been responsible for destroying seed stores in Rwanda, Burundi, and the Solomon Islands.

In 2004, Iraq's seed bank in the now infamous Baghdad suburb of Abu Ghraib was looted during the insurgency. Iraqi scientists had been savvy enough to send copies of its seed stock across the border to the International Center for Agricultural Research in the Dry Areas (ICARDA), a CGIAR center in Aleppo, Syria. However, ICARDA subsequently had to safeguard its own extensive seed collection against the ongoing hostilities in Syria by sending duplicates to Svalbard Global Seed Vault. Of critical importance, the majority of ICARDA's 150,000 accessions are irreplaceable landraces and wild crop relatives of plants gathered from locations where crop domestication—and civilization—first emerged.

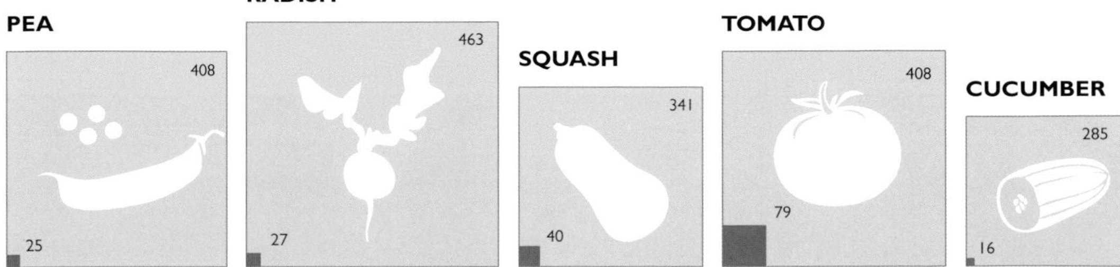

1 *Svalbard Global Seed Vault is located deep inside a mountain on an island of the Svalbard archipelago. Even if the power fails, the seed samples it stores will remain frozen.*

2 *Parcels of seeds are transported through a blizzard to the seed vault.*

3 *Seeds are taken from the entrance along a 330 feet (100 m) long tunnel to the main chamber.*

4 *The black nightshade* (Solanum nigrum), *one of the species of the Solanaceae family. The Solanaceae Germplasm Bank specializes in storing seeds of wild non-tuber-bearing nightshades.*

COMMUNITY SEED BANKS SUPPORT FARMING COMMUNITIES

Alongside the larger official seed banks, many smaller community banks have sprung up around the world, often aided by non-governmental organizations (NGOs). The motivations behind their establishment range from biodiversity loss and natural disasters, to climate change and declining access to good-quality seeds. Although varying greatly in size, these seed banks tend to have a single common goal: to maintain varieties of seeds for use by local farming communities. Local-level institutions enable people who might not have access to more formal seed collections to conserve their own varieties and landraces, and gain help with selecting, producing, improving, and marketing seeds and crops.

WILD SPECIES AT HEART

An increasing number of seed banks are concentrating on preserving the biodiversity of wild plants. The seed bank of the Universidad Politécnica de Madrid (UPM), in Spain, was a pioneer in banking seeds of wild species. Established in 1966, it uses ex situ methods to conserve wild taxa and makes them available for research purposes. It focuses on saving plants from the Cruciferae (Brassicaceae) family, as well as rare and endangered species that grow in the western Mediterranean region.

The Solanaceae Germplasm Bank, meanwhile, at the Botanical Garden of Nijmegen, the Netherlands, holds a vast collection of non-tuber-bearing Solanaceae, or nightshades, mostly gathered from the wild. Its aim is to conserve the genetic diversity within the Solanaceae, an economically valuable plant family encompassing crops, spices, medicinal plants, and ornamentals.

Kew's MSBP is unique in the extent of its ambitions and network, collaborating with partner organizations in more than 80 countries. Having concentrated in its early years on gathering seeds and working out how to germinate them, the MSBP's focus is now shifting to putting seeds to use to help manage and maintain biodiversity in the landscape. "I think we're on the cusp of seed banks really realizing their full potential," says Paul Smith, former Director of the MSBP and now Secretary General of Botanic Gardens Conservation International. "If you want resilient plants, you need diversity in the landscape. Seed banks can supply a full range of diversity, along with information on how to grow that in the real world."

The need to reverse biodiversity loss was starkly highlighted by the Millennium Ecosystem Assessment, initiated in 2001. Called for by the then Secretary-General of the United Nations, Kofi Annan, the Millennium Ecosystem Assessment aimed to assess the consequences to human well-being of ecosystem changes, and to provide a scientific basis for action to enhance the conservation and sustainable use of those systems. The 1360 experts who contributed to the assessment came to the grave conclusion that, "Over the past 50 years, humans have changed ecosystems more rapidly and extensively than in any comparable period of time in human history, largely to meet rapidly growing demands for food, fresh water, timber, fiber, and fuel. This has resulted in a substantial and largely irreversible loss in the diversity of life on Earth."

HEALTHY ECOSYSTEMS ARE KEY TO BIODIVERSITY

Today, the once disparate aims of agricultural and wild species seed banks, to underpin food security and to conserve biodiversity, are being united by the need to conserve functioning ecosystems. Scientists now understand that conserving biodiversity by simply saving individual cultivars or species is not sufficient. This is because plants, whether cultivated or wild, form a vital part of functioning ecosystems in which they interact with animals, soils, and the atmosphere. Humans gain great benefits from different kinds of healthy ecosystems. For example, mangrove swamps protect cities from floods; rainforests extract carbon dioxide from the air and recycle water; and flower-rich woodlands and meadows support bees and other insects that pollinate farmers' crops. These services are becoming weakened through widespread environmental degradation. To ensure our continued future use of them, we need to conserve biodiversity and shore up entire ecosystems.

An example of how maintaining natural ecosystems can be beneficial to agriculture comes from Costa Rica. Scientists who studied populations of wild bees living near a single coffee farm found that patches of biodiversity around the farm's fields nurtured bee populations, which, in turn, promoted effective pollination. Pollination is fundamental to farming, as many crops—including almonds, avocados, cherries, and coffee—require this natural service in order to reproduce. Bees, which provide the majority of crop pollination, generally only fly half a mile (0.8 km) or so before needing to feed on flower nectar. In the Costa Rica study, the researchers found that coffee yields were higher by 20 percent in areas of the farm that lay within half a mile of biodiverse forest patches.

BIODIVERSITY RESTRICTS THE SPREAD OF DISEASE

Another service provided by ecosystems is that of buffering the spread of infectious diseases. A regional genotype of hantavirus causes hantavirus pulmonary syndrome (HPS), a severe respiratory disease in the Americas, with some 637 cases of HPS having been reported in the USA by the end of 2013. The virus is transmitted by rodents, which pass it to other animals of the same species during aggressive encounters. Humans can catch it when they come into contact with infected animals or their excreta. Scientists noted that recent hantavirus outbreaks, which were transmitted from host animals to humans, occurred in habitats that had been highly disturbed by human activity, reducing their biodiversity.

Evidence suggested that a "generalist" rodent species was responsible for spreading the disease. Generalists survive in a wide range of habitats and can eat a range of foods. They tend to invest little in immune defense, however, which makes them more vulnerable to pathogens. Specialist species, on the other hand, occupy a niche habitat and rely on only a few food sources. They tend to invest more in their immune systems. When the biodiversity of specialist species declines within a given ecosystem—for example, when disturbed by human activity—a generalist species is able to move in and quickly multiply. With fewer specialist species around, the generalist incomer will encounter animals of its own species more frequently, enabling it to pass on the hantavirus more effectively.

RETURNING LOST SPECIES

As scientists have become increasingly aware of the importance of biodiversity and the interactions within ecosystems, they have begun to examine ways of restoring degraded habitats. For example, patches

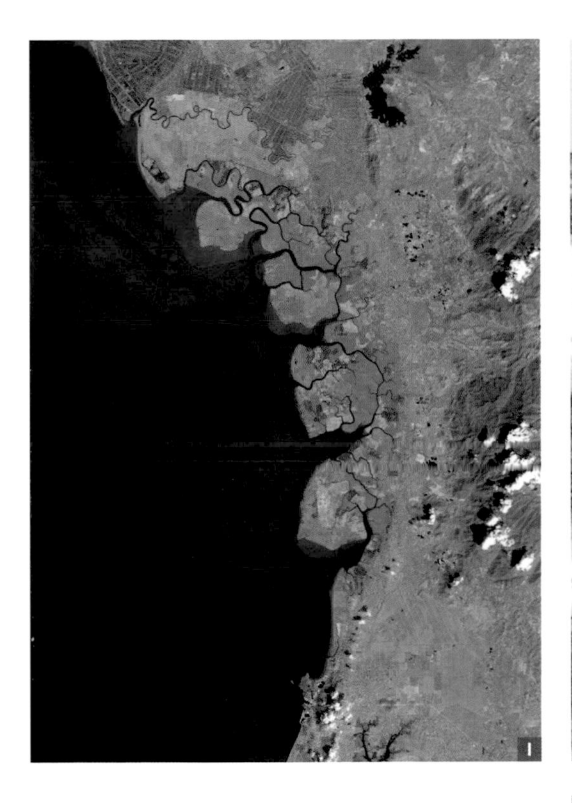

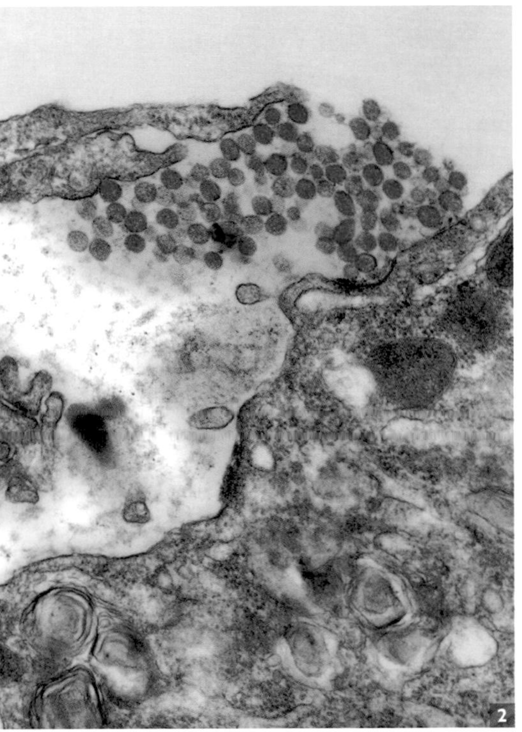

of South Africa's critically endangered Cape Flats sand fynbos habitat, which has declined by more than 90 percent, have been rejuvenated. By returning some key missing plant species to a relatively healthy patch of the habitat at Kenilworth racecourse, Cape Town, scientists have attracted monkey beetles, sugarbirds, and Cape sunbirds back to the area for the first time in 100 years. The "Purcell List" is an inventory of plants encountered on the Cape Flats by amateur botanist William Frederick Purcell during the eighteenth century. The scientists are using his observations to build up seed stocks of all the plants represented in the Cape Flats sand fynbos habitat, so they can restore other areas.

1 *Mangrove swamps ameliorate the impact of floods, reducing damage to inland towns and cities. This satellite image shows Matang Mangrove Forest, a sustainably managed area on the coast of Malaysia. The mangroves are dark green; agricultural fields (rectangular) are light green; and urban or cleared areas are pink.*

2 *A colored transmission electron micrograph of hantavirus particles (blue) bursting from a human cell. The virus, which infects humans when they inhale or ingest excreta from infected rodents, causes hantavirus pulmonary syndrome and hemorrhagic fever with renal syndrome.*

RESTORING BIODIVERSITY TO CHALK DOWNLANDS

Kew's MSBP is building up stocks of seeds for restoring degraded chalk downland habitats, under its UK Native Seed Hub Initiative. Not far from the undulating building that houses the seed vault is a field in which 32 beds, each 65 feet (20 m) long, are growing a range of chalk downland species, specifically for their seeds.

Chalk grasslands typically have low levels of nutrients, and experience summer droughts and winter frosts. These conditions, when combined with grazing, create a habitat in which no one species reaches its full potential. Instead, many species flourish alongside each other, forming a very biodiverse habitat. As many as 40 species of plants can grow on one square yard (meter) of chalk. Such pastures thrived in southern England as recently as 50 years ago but they began disappearing with the expansion of mechanical farming, the loss of grazing by sheep, and the introduction of fertilizers and pesticides.

In 2010, the UK Government published the results of an independent review of England's wildlife sites led by the eminent British ecologist John Lawton. Called *Making Space for Nature*, the review assessed wildlife sites with a view to seeing if they would be able to adapt to new conditions brought about by climate change. It concluded that England's varied wildlife sites were generally too small and too isolated to adapt easily, which would almost certainly lead to declines in many of England's characteristic species. The report suggested that the situation was likely to get worse with climate change. It recommended better protection and

1 *Devil's bit scabious (*Succisa pratensis*), which thrives in damp meadows and marshes, being grown for the UK Native Seed Hub project. The species is a food plant of the Marsh Fritillary butterfly, a declining UK species.*

2 *Sheepcote Valley, a nature reserve close to Brighton, UK. The South Downs National Park NIA aims to improve grassland biodiversity here and at other sites.*

management for designated wildlife sites, protection for non-designated wildlife sites, and the formation of new ecological restoration zones.

A CENTER FOR SEED EXPERTISE

The UK Native Seed Hub was set up in 2011, with funding from the Esmée Fairbairn Foundation. The idea was to bulk up supplies of seeds of native English plants and then to supply them to commercial seed organizations, conservation organizations, botanical charities and other authorities engaged in restoring habitats. "We wanted to contribute to the vision encapsulated in the Lawton review, and subsequent policy documents, of strengthening the ecological network in the UK," explains Ted Chapman, Coordinator of the UK Native Seed Hub. "For active habitat restoration and creation, you need high-quality plant materials. That's the niche that the UK Native Seed Hub fills. As well as providing plant materials, we offer technical support in the areas we're good at, such as seed testing, seed storage, post-harvest handling of seeds, and so on."

NATURE IMPROVEMENT ON THE SOUTH DOWNS

The UK Government's 2011 Natural Environment White Paper followed up on the Lawton review' s recommendations by introducing the concept of Nature Improvement Areas (NIAs). These were to receive funding so authorities could revitalize urban and rural areas by creating networks of wildlife habitats. Twelve NIAs were selected, including the South Downs National Park, the vision for which is a "better connected and inspirational chalk ecosystem, sustainably managed to enhance biodiversity and people's well-being for now and the future". The UK Native Seed Hub is contributing directly to this goal through a partnership with Brighton and Hove Council. Early on in the project, MSBP staff brush-harvested seed (a method of gathering large quantities of seed quickly) from one of the Council's chalk grassland sites, Bevendean Down. Half the seeds went to the MSBP for generating more seed for the Seed Hub's plots, and the remainder went to the Council for its NIA work.

On a rainy winter's day, inside the modest single-story red-brick offices of Brighton and Hove Council's Stanmer Nurseries, within Stanmer Park, Conservation Manager David Larkin explains about the work being undertaken to fulfill the South Downs NIA's objectives. The Council and MSBP are among 29 organizations, led by the South Downs National Park Authority, involved in the initiative. Stanmer Nurseries' role is to help nurture the wildflowers that grow in chalk downlands, while in turn attracting bees and butterflies. "For centuries farmers grew corn in the valley bottoms of the South Downs and

grazed sheep on the low-fertility hills," explains David. "The sheep were folded at night on fallow fields, so over time there was a transfer of nutrients from the already low-fertility hills onto the arable land. Hundreds of years of grazing and lowering nutrient levels produced this really ecologically rich chalk grassland."

Over time, this traditional farming practice changed. With the invention of refrigeration in the 1870s, lamb could be imported from New Zealand, so sheep farming on the South Downs began to decline. The development of a process to synthesize nitrate from the air led to chemical fertilizers, which sounded the industry's death knell. Farmers could now improve soils in the valley bottoms and on the hillsides alike, without sheep dung. Over the past 60 years, virtually all the old grasslands have been ploughed up to grow corn. "Now there's only three percent of ancient grassland left," says David. "That's why we're trying to conserve the chalk grassland. The main priority is to not lose any more and the second priority is to try to expand and join up the bits that are left."

SEEDLINGS AND SHEEP TO SAVE GRASSLAND

Inside one of several glasshouses lined up opposite the Nurseries' offices are hundreds of trays of diminutive plants. A single tray holds 66 plants, with each tiny green rosette sitting in its own plug of soil. Among the seedlings are bird's-foot trefoil (*Lotus corniculatus*), cowslips (*Primula veris*), bladder campion (*Silene vulgaris*), and salad burnet (*Sanguisorba minor*), all of which grow on chalk downlands. Grown by semi-retired nurseryman John Gapper and a team of volunteers, using the MSBP's brush-harvested seed and other locally made collections, they are among 90,000 plug plants raised by the nursery each year for transfer to local sites. "John has worked for the Council for 40 years," says David. "He has an interest in wild flowers and has spent his own time trying to propagate them. When the NIA project came along we were able to build on his expertise."

The plants are being transferred to various sites located in and around the city of Brighton. The largest site, some 300 acres (120 hectares), incorporates East Brighton Park and Sheepcote Valley. Standing beside a newly installed fence at the southeastern end of Sheepcote Valley, David explains that scrub has encroached on the valley bottom, which is now an important feeding ground for migratory birds. The valley sides, however, remain as grassland.

The Council has fenced the land and brought its flock of Herdwick sheep to graze the area, in an attempt to enhance the grassland biodiversity. The sheep are regularly shifted between the Council's different ecological sites by local farmer Sam Baldock in order to emulate bygone farming practices. A group of 120 volunteer shepherds take turns keeping a watchful

1 *Salad burnet (*Sanguisorba minor*), one of the species being used to repopulate chalk downland in East Sussex.*

2 *Wild carrot (*Daucus carota*). Unlike its edible cousin, the wild carrot's root is tough and unpalatable.*

3 *The large blue butterfly (*Maculinea arion*) initially proved hard to conserve because researchers were ignorant of its association with the red ant (*Myrmica sabuleti*).*

eye over them each day, to ensure they obtain sufficient nutrients from the unimproved grassland to remain healthy.

Looking west into East Brighton Park, a large white circle, around 160 feet (about 50 m) in diameter in a "yin-yang" shape, clearly stands out from the green of the surrounding grass. This is one of the repositories for John Gapper and the volunteers' plug plants. After removing the topsoil, the team planted 11,000 plants directly into the chalk in the hope that a species-rich chalk grassland will emerge in the years to come. Tiny plants of wild carrot (*Daucus carota*), sheep sorrel (*Rumex acetosella*), devil's-bit scabious (*Succisa pratensis*), and salad burnet (*Sanguisorba minor*) are already thriving within the creamy white bedrock. "We've enhanced some 12 to 13 sites so far in this way and we have 15 in total to do for the NIA," explains David. "The closer the fragments of quality chalk grassland are to each other, the more likely they are to attract butterflies and the greater chance we have of increasing the biodiversity of the ecosystem."

The three-year NIA project is now drawing to a close, but it is likely to be several years before David and his team are able to see if their work has been successful in creating functioning chalk grassland habitats from fragments of the once widespread ecosystem. "There are some very complex interrelationships in the chalk grassland web, and no one understands them all fully," he admits. "For that reason, we can't necessarily predict what a restored site will look like. Different sites have had different management and have different aspects and soils. Some plants are always going to favor some sites over others. So when we put in plug plants we generally plant a whole range of things in the hope that a good number of them will survive."

BUTTERFLIES DEPEND ON ANTS

The difficulties of trying to restore habitats are exemplified by the case of the large blue (*Maculinea arion*) butterfly. Once native to England, the species became extinct in the British Isles in 1979. The conservationists who were trying to save it didn't realize at the time that its caterpillars form an association with the red ant (*Myrmica sabuleti*). The pupa of the large blue produces chemicals that trick worker ants into thinking it is ant larvae, prompting the ants to take it into their nest. Once there, the caterpillar mimics the sound of the queen ant so faithfully that worker ants care for it. Here it survives the winter, feeding on ant larvae. "The conservationists were not conserving the site in a way that nurtured the ants, so the ants disappeared and the butterflies followed suit," says David. "Now the relationship is understood and the large blue butterfly has been successfully reintroduced from populations in Sweden." By 2006, an estimated 10,000 adult large blue butterflies were occupying 11 British sites, a conservation success story.

THE SHIFTING STATE OF THE WORLD'S FLORA

1799
German naturalist and explorer Baron Alexander von Humboldt expresses his concern about the rate at which cinchona trees are being felled in South America. At this time, cinchona bark is being used to treat malaria.

1804
The world's population reaches one billion.

1927
The world's population reaches two billion (having taken 123 years to increase by a billion).

1943
The world is at war. Recognizing the "botanical intelligence" that seeds represent, an elite Nazi commando squad raids scientific institutes in Ukraine and the Crimea to obtain seeds, collected through the inspiration of Nikolai Vavilov, in one of the biggest acts of biopiracy in history.

1945
With the end of the war comes an emerging awareness of environmental degradation.

1960
The world's population reaches three billion (having taken 33 years to increase by a billion).

1970
The first list of threatened plants is produced by the newly formed International Union for the Conservation of Nature (IUCN). It suggests that 20,000 species of plants need some protection to survive.

1972
The United Nations Conference on the Human Environment is held in Stockholm, Sweden.

1974
The world's population reaches four billion (having taken 14 years to increase by a billion).

1975
The Convention on International Trade in Endangered Species (CITES) enters into force, designed to ensure that international trade in plants and animals does not threaten their survival in the wild.

1987
The world's population reaches five billion (having taken 13 years to increase by a billion).

1989
Henry Azadehdel becomes the first person in the UK to be sentenced to a year in prison for orchid smuggling. However his sentence is reduced on appeal to five weeks. He is also fined $15,200 (£10,000) (later reduced to $3,800 [£2,500]).

1992
The first International Earth Summit takes place in Rio de Janeiro, Brazil. The meeting concludes that attitudes and behavior must be transformed to reverse damaging stresses being placed on the environment by both poverty and overconsumption. The conference gives rise to Agenda 21, the Rio Declaration on Environment and Development; the Statement of Forest Principles; the United Nations (UN) Framework Convention on Climate Change and the UN Convention on Biological Diversity.

1999
The world's population reaches six billion (having taken 12 years to increase by a billion).

2000
The Millennium Summit of the United Nations in New York agrees on eight Millennium Development Goals, to be achieved by 2015. They include the goal to "ensure environmental sustainability" by, among other processes, "reducing biodiversity loss" and "integrating the principles of sustainable development into country policies."

The Millennium Seed Bank Project (now Partnership) (MSBP) opens at Wakehurst Place, West Sussex, UK. Opening the seed bank, Prince Charles describes it as "the Bank of England of the botanical world."

2008

The Svalbard Global Seed Vault opens 800 miles (1290 km) above the Arctic Circle. It is the world's largest secure seed vault. Its primary aim is to preserve the vast genetic diversity present in crops around the world.

The plant species *Crepis sancta* is found to produce heavier seeds in urban areas than when inhabiting rural locations. Scientists suggest that having heavier seeds that fall close to the mother plant would prevent the seeds landing on concrete. But they also conclude that plants which adapt to cities in this way, far from ensuring the survival of their species, may even increase their chances of extinction.

2009

Ahead of schedule, the MSBP achieves its goal to bank seeds from 10 percent of the world's flora. Its work continues, with the aim of banking seeds from 25 percent of the world's flora by 2020.

2010

Research undertaken jointly by Kew Gardens, the Natural History Museum, and IUCN finds that a fifth of the world's plant species are at risk of extinction. Of 4,000 species examined by the Sampled Red List Index for Plants, 22 percent are considered to be "threatened," while 33 percent are too poorly understood to be assessed.

The UK's pavilion at the Shanghai Expo highlights the importance of biodiversity. The "Seed Cathedral," designed by renowned architect–designer Thomas Heatherwick, comprises a 50 feet (15 m) high wooden cube pierced with 60,000 25-feet- (7.5-m-) clear acrylic rods. The end of each rod encloses one or more seeds. The aim of the exhibit is to encourage visitors to reflect on the importance of conserving natural resources by adopting more sustainable lifestyles, particularly in cities.

2011

The MSBP launches the UK Native Seed Hub, to help restore and protect the UK's wildflower meadows. Only two percent of meadows that existed at the end of the Second World War in 1945 still remain intact.

The world's population reaches seven billion (having taken 12 years to increase by a billion).

Studies of thale cress (*Arabidopsis thaliana*) reveal the gene behind seed germination. This discovery could help scientists develop crops that are drought- and flood-resistant.

A new species of plant is discovered, which "bends down" to bury its own seeds. The plant, from the Atlantic forest of Bahia, northeastern Brazil, is named *Spigelia genuflexa*, after its unusual seed-dispersal mechanism. Scientists still have much to learn about the many adaptations that help plants survive.

2012

Twenty years on from the Earth Summit, the Rio+20 United Nations Conference on Sustainable Development takes place in Brazil. Member states agree to develop Sustainable Development Goals, building upon the Millennium Development Goals.

2013

Nature Climate Change reports that more than half of common plant species (along with a third of animals) face a serious decline in their habitat range due to climate change if rapid action is not taken to cut greenhouse gas emissions.

Scientists at Doñana Biological Station in Seville, Spain, discover that the destruction of the Brazilian rainforest is causing seeds to become smaller and weaker, giving them less chance of survival. They blame the disappearance of large fruit-eating birds; with these species absent only smaller birds exist to disperse the seeds, meaning only the smaller seeds digestible by those species will be spread.

The MSBP establishes the UK's first national collection of tree seeds, comprising 77 species.

2014

A new species of "nickel-eating" plant, named *Rinorea niccolifera*, is discovered on Luzon, the largest island in the Philippines. It could potentially be used to help clean up polluted ecosystems. There are many as yet unnamed and uncataloged plants that may prove to have traits of benefit to humanity.

2015

A study finds that commonly prescribed drugs, such as ibuprofen and diclofenac, can adversely affect the growth of edible crops, including lettuce and radish, even at the low concentrations found in the environment.

The Svalbard Global Seed Vault accepts its first consignment of tree seeds, aimed at helping to monitor long-term genetic changes in forests.

A study finds that even moderate climate change will affect India's endemic plants but suggests that a program of expanding protected areas and helping plants migrate to them could help save them.

The US National Academy of Sciences says climate change has already advanced so much that interventions on a planetary scale are needed.

SEED BANK SOUTH AMERICA

KEEPING HUNGER AT BAY IN THE TROPICS

NAME

International Center for Tropical Agriculture (CIAT)

NUMBER OF ACCESSIONS

CIAT's Genetic Resources Program holds the largest genetic collection of beans (37,000 accessions), tropical forages (23,000 accessions), and cassava (7,000 accessions) in the world.

WHEN FOUNDED

CIAT was founded in 1967. It is one of 15 international agricultural research centers that make up the CGIAR Consortium.

FOCUS OF THE COLLECTION

The beans and forages collections are primarily banked seeds, while the cassava collection is held as tissue samples. The beans comprise 45 species of the genus *Phaseolus* from 110 countries, while the forages include 734 species of grasses and legumes from 75 countries. The majority of the bean collection is made up of cultivated material, but there are around 2,000 accessions from wild plants. The cassava accessions, gathered from 28 countries, comprise around 6,000 clones of *Manihot esculenta* and 900 genotypes of wild species. "Since its foundation, CIAT has distributed more than half a million samples to farmers, breeders, and scientists," explains

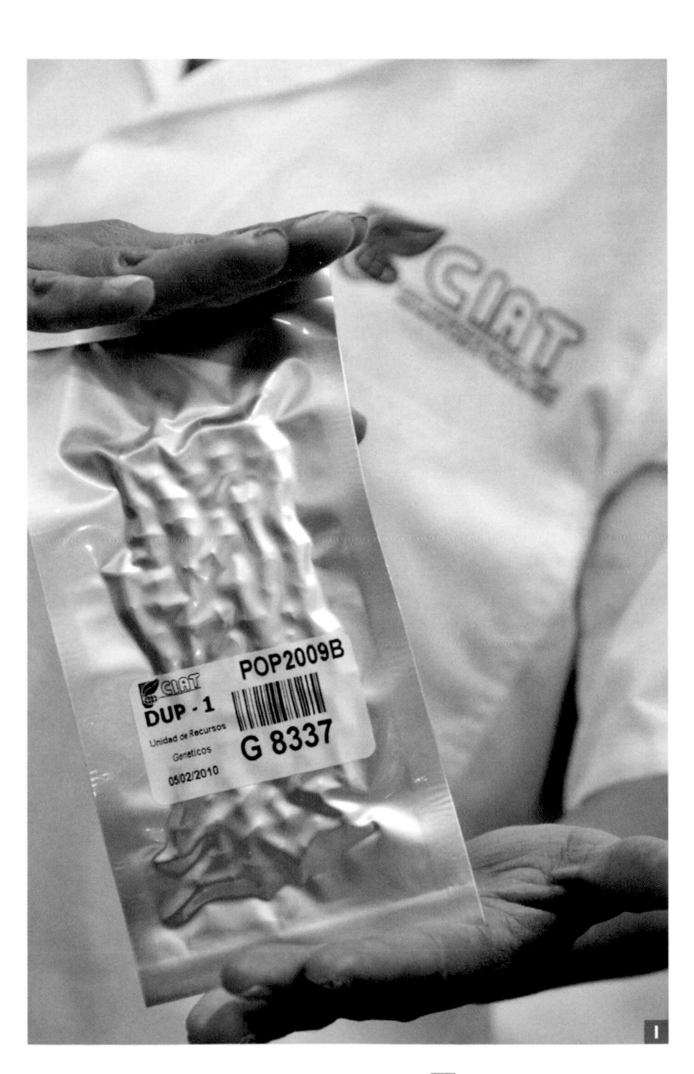

1 *This vacuum-packed bag of seeds is ready for long-term storage at CIAT.*

2 *Protective clothing is vital in the coldest vault of the CIAT gene bank, where the temperature is -4°F (-20°C).*

3 *Barcodes are used to denote different bean varieties.*

Daniel Debouck, manager of CIAT's gene bank. "We are the sole provider of many bean and forage seeds. CIAT's latest review recommended that we increase our stocks of beans and forages, so we're currently involved with expanding those stocks. In a couple of years from now the bean collection will hold around 45,000 accessions and the forages should comprise around 25,000 accessions."

WHY IT IS NEEDED

CIAT aims to reduce hunger, cut poverty, and improve nutrition in the tropics by increasing the eco-efficiency of agriculture. Genetic resources, which can be used to breed crops that, say, require less water or are more resistant to disease, are fundamental to achieving these goals.

WHO FUNDS IT

A major contribution comes from CGIAR. Additional grants are provided by the Global Crop Diversity Trust.

WHERE SEEDS ARE STORED

All the genetic material is held at CIAT's headquarters in Colombia. Seeds reside in two major cold stores, one kept at a temperature of –4°F (–20°C) and the other at 41°F. (5°C). There are also facilities for storing *in vitro* collections, whereby plants are conserved as cells, tissues, and organs. Duplicate accessions are held at the Svalbard Global Seed Vault and at the International Maize and Wheat Improvement Center, of CGIAR, in Mexico City.

CURRENT RESEARCH

"Our research has always been designed to answer two major questions," explains Daniel. "One is: 'How can we improve conservation protocols, so that we can conserve material in a way that it will endure?' We need to make sure we have sufficient accessions over the long term for us to send material out in response to requests. The other question is: 'What should be in the collection to ensure it is as relevant as possible?' It's important to make sure we have useful genetic diversity in the collection."

The findings of a project demonstrating the usefulness of the collection were published in 2015. CIAT scientists identified 30 "elite" lines of beans that can tolerate temperatures 7°F (nearly 4°C) higher than current crops' favored conditions. Many of these tolerant lines came from crossing the widely grown common bean *(Phaseolus vulgaris)* with a lesser-known relative called the Tepary bean *(Phaseolus acutifolius)*. This bean was domesticated in the arid climate of southwestern USA and northern Mexico. As a result, it is more able to thrive under dry, hot conditions than any other grain legume.

By analyzing 19 models showing potential future climates, the scientists concluded that the area suitable for growing today's bean cultivars could have shrunk by up to half by 2050. When the current cultivars were replaced by the heat-tolerant beans, however, the production area was forecast to reduce by only 5 percent. The new lines therefore represent a buffer, which could help ensure poor populations in tropical nations have food to eat under warmer climatic conditions.

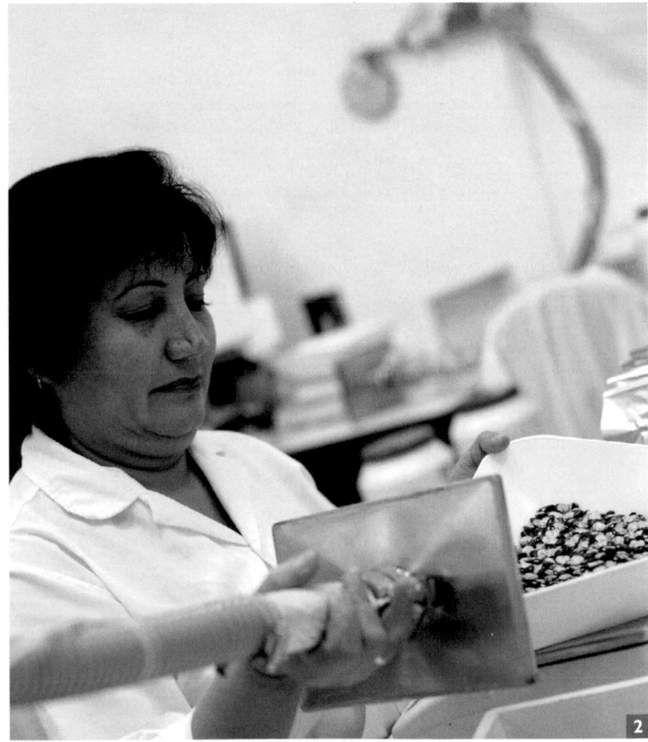

1 *Cassava growing* in vitro. *Cassava is vegetatively propagated, so rather than storing its seeds, the scientists at CIAT grow cuttings in growth media under controlled conditions.*

2 *This Hoover-like tool enables scientists to suck up and move large numbers of beans at once.*

3 *Beans are carefully labeled for easy recognition.*

4 *A researcher prepares to plant bean seeds at CIAT's headquarters in Colombia.*

SEEDS WITH A STORY

"When I was at Kew a few years ago, I came across a sample of a bean that grows in the Bermuda Archipelago," recalls Daniel. "I noticed that seeds of the bean, *Phaseolus lignosus* Britton were labeled as being conserved at the MSBP, so I contacted the seed bank and asked if they could send me a few. In due course they sent me 10 seeds, which I subsequently increased up to some 6,000 seeds. This was much easier for me to do in Colombia's climate than for the MSBP to do in the UK's climate.

In Bermuda, there were only between three and five populations left of that particular species, and we only had seeds from a single population. Having been able to increase the number of plants and seeds available, we now have the possibility to restore the plant's habitats in Bermuda by sending seeds back to the island.

We later discovered that the bean is a wild relative of the edible Lima bean (also known as the butter bean [*Phaseolus lunatus*]), which is cultivated in tropical and semitropical nations around the world as a source of dietary fiber and protein. So, it's possible that, in future, genetic material from the Bermuda bean could help to make the lima bean more resilient to climate shifts."

FOODS OF THE FUTURE

Although 7,000 food crops are cultivated or harvested from the wild, we derive half our plant-based calories from just three: rice, maize, and wheat. Sorghum, millet, potatoes, sweet potatoes, soybean, and sugar (from cane and beet) account for a further 25 percent of our energy intake. With some widely used cultivars potentially vulnerable to climate change, the search is on for new resilient crops. Here are some presently under-used crops that could contribute to feeding the world in future.

HULLED WHEATS (*TRITICUM SPECIES*)

Hulled wheats were among the earliest plants to be domesticated and were staple crops for many millennia. The three main domesticated hulled wheats are einkorn, emmer, and spelt. Hulled wheats have a protective layer, absent in commonly used bread and durum wheats, which remains intact when harvested and has to be removed before milling. The ancient wheats dwindled in importance due, in part, to competition from new crops that were easier to process. However, hulled wheats have a wide array of uses including making bread, porridge, gruel, soup, cracked wheat, and beer. They are now making a comeback among foodies in Europe.

JACKFRUIT (*ARTOCARPUS HETEROPHYLLUS*)

Despite having a long history as a staple food in India, the jackfruit is now shunned on the subcontinent as a "poor person's fruit." Its lowly reputation is unwarranted, as jackfruits are highly nutritious, containing potassium, calcium, and iron. The tree is easy to grow and tolerates high temperatures and drought. Efforts are now under way to try and reverse the jackfruit's bad reputation and expand its production across India. The fruit is enormous: a single fruit can weigh as much as 100 lb (45 kg). With this impressive yield, and climate change already reducing harvests of wheat and maize, the lowly jackfruit has the potential to become a king of foods.

BAMBARA GROUNDNUT (*VIGNA SUBTERRANEA*)

The Bambara groundnut is a highly nutritious legume that is mainly grown by women farmers as a subsistence crop in drier parts of sub-Saharan Africa. Its seeds, which grow underground, can be boiled, roasted, or fried, as well as used to make milk and flour. The Bambara groundnut provides relatively good yields on poor soils with little rainfall and significantly greater yields under better conditions. However, it is still grown largely using local farmer-derived landraces rather than varieties that have been specifically bred for particular agricultural conditions. Seeds from wild relative plants of the Bambara groundnut are being collected as part of the Adapting Agriculture to Climate Change project being undertaken by the MSBP and the Global Crop Diversity Trust (highlighted in chapter 1). The project will make the seeds available to breeders so they can develop new varieties that are resilient to climate change.

AMARANTH (*AMARANTHUS SPECIES [A. CAUDATUS, A. CRUENTUS, AND A. HYPOCHONDRIACHUS]*)

The pre-Columbian Aztecs ate the seeds of the Amaranth plant as a staple crop but were forbidden from growing it by the invading Spanish conquistadors. Today it is hailed as a "superfood" because of its nutritional properties; it is a complete protein containing all the essential amino acids, is gluten-free, and has been found to lower cholesterol.

FINGER MILLET (*ELEUSINE CORACANA*)

Finger millet is a small, grassy plant with highly nutritious seeds and the ability to survive severe drought. It is cultivated in 25 countries across Africa and Asia as a staple food grain, but its use has been declining because of the crop's low productivity. Organizations such as the International Crop Research Institute for the Semi-Arid Tropics (of CGIAR) and the Kenya Agricultural Research Institute are helping to produce new varieties with higher yields. The crop's ability to withstand drought, and for its seeds to be stored for long periods without insect damage, make it a reliable famine food to sustain people during hard times.

TEFF (*ERAGROSTIS TEF*)

In Ethiopia, the small millet known as teff accounts for 20 percent of all land under cultivation. Its seeds, which are tiny like poppy seeds, are high in calcium, iron, and protein. Also gluten-free, they can be ground to make a substitute for wheat flour. Teff products are gaining popularity among Western consumers of health foods. A lack of teff on international markets has led farmers to begin growing it in the USA, Europe, Israel, and Australia. Whether Ethiopian farmers can benefit from the crop's newfound middle-class consumers remains to be seen.

BRINGING THE ORPHANS INTO THE FOLD

Crops that are locally significant but under-researched and under-funded are called "orphan crops." Many are indigenous to Africa, Asia, and Latin America. They are often highly resilient to environmental stresses, such as droughts, but have limited importance in the global marketplace. With climate change now evident around the world, greater attention is being paid to such crops. Technological developments, such as creating new hybrids, have the potential to greatly improve the performance of orphan crops. For example, the productivity of teff in Ethiopia has increased by 5 percent over the past 16 years, much more than that of wheat or maize over the same time period.

*Finger millet (*Eleusine coracana*) is a drought tolerant crop with nutritious seeds. Long relied on as a famine food in Africa, it has potential to become more widely used thanks to new, higher-yielding varieties.*

PLANTS AND SEEDS FROM THE WORLD'S ALPINE HABITATS

The term "alpine" refers to plants that naturally grow in mountainous areas of the world, above the altitude at which trees can survive. Frequently low-growing, with seeds primarily dispersed by the wind, they have adaptations that enable them to tolerate strong winds, low temperatures, snow cover, and high levels of ultraviolet light. Most alpine floras have a high number of endemic species. The highest level of endemism occurs in the Himalayas, where around half of the plants are not found anywhere else. Scientists estimate that between 30 and 50 percent of alpine plants are threatened by climate change, making them a high priority for seed-banking. Here is a selection of some at-risk species.

SAXIFRAGA TOMBEANENSIS

This saxifrage is native to the Italian Alps, where it grows at altitudes between 2,300 and 7,000 feet (700 and 2,100 m). It is one of 350 plant species, with very specific habitat requirements, that live in the Alps and nowhere else. Saxifrages are adapted for dispersing seeds in winter; they have stiff flower stems that can withstand snow and strong winds. Their seeds are small and round, so they can easily roll away from the plant under gravity in mountainous environments. The name saxifrage comes from the Latin *saxum* (a rock) and *frangere* (to break), because its presence on rock faces suggested it was capable of breaking rocks.

Saxifraga tombeanensis forms low cushions on limestone and dolomite vertical cliffs. Its low level of seed germination, coupled with collection by botanists and gardeners, and encroaching woodland species such as *Ostrya carpinifolia*, *Pinus sylvestris*, and *P. mugo*, are putting pressure on the species. In 2010, it was listed as endangered by the IUCN. Climate change is likely to put further pressure on the plant. Since the 1980s, meteorological stations close to its distribution range have recorded a strong increase in mean annual temperature. Seeds of *Saxifraga tombeanensis* are stored in seed banks in Lombardy and Trentino-Alto Adige, as an insurance policy against the plant's extinction.

PODOCARPUS NIVALIS

The snow totara (*Podocarpus nivalis*) grows as a creeper or low bush at altitudes between 2,600 and 8,200 feet (about 800 and 2,500 m) in alpine and subalpine environments of New Zealand. Distributed widely on North Island, South Island, and Stewart Island, it is currently listed by the IUCN as being of "Least Concern." However, overgrazing has caused some populations to decline.

Scientists recently discovered that seeds of the snow totara are among those dispersed by the New Zealand kea (*Nestor notabilis*), the world's only alpine parrot. The fruits of *P. nivalis* were among those favored by the kea, constituting 60 percent of observed feeding time at two study sites. Many other parrots destroy most of the seeds they eat with their beaks. However, of 219 seeds of *P. nivalis* extracted from kea faecal samples, the majority (99.1 percent) were intact.

In total, the scientists found that 19 seed species were present in the kea faeces, mostly undamaged. They concluded that the alpine parrot is the most numerically dominant avian seed disperser for most fruiting species in New Zealand's alpine ecosystems. Kea populations have declined drastically over the last 120 years; fewer than 5,000 now exist in the wild. If their numbers are further reduced there may be implications for New Zealand's alpine plants, including *P. nivalis*.

SEEDS THAT LIVE THE HIGH LIFE AGE FASTER

Seed scientists at Italy's University of Pavia and the MSBP have found that seeds from plants growing in colder, wetter habitats at higher altitudes do not live as long as their counterparts in warmer, drier lowland settings. The scientists collected seeds from 69 plant species growing in alpine and lowland locations across Italy. They exposed the seed lots to a temperature of 113°F (45°C) and 60 percent relative humidity, to encourage them to grow, and recorded when seeds germinated. They calculated the time taken for the viability of a particular seed lot to fall to 50 percent.

The findings showed that seeds were progressively longer-lived with increased temperature and decreased rainfall. Plants growing in alpine habitats are particularly sensitive to land-use change (such as tourism developments) as well as climate change, so storing their seeds is important for conservation. The conclusions suggest that seed banks will have to test regularly the ability of their stored alpine species to germinate, to ensure that seeds are still viable. More research is needed to find out the extent to which seed longevity is determined by genetic or environmental influences.

*The seeds of the snow totara (*Podocarpus nivalis*) are dispersed by the New Zealand kea (*Nestor notabilis*). However, fewer than 5,000 individuals of the kea now exist in the wild.*

RHEUM EMODI

Rheum emodi (Himalayan rhubarb) is a leafy perennial herb that grows in subtropical and temperate parts of the Himalayas at altitudes between 8,200 and 14,750 feet (2,500 and 4,500 m). It has a long history as a medicinal plant, and was a major export from Asia by the tenth century. Traditionally used to treat fevers, coughs, indigestion, and menstrual disorders, it has been found to have anticancer, antioxidant, antidiabetic, antifungal, and antiulcer properties.

Overharvesting of the rhubarb for medicinal purposes, together with poor seed germination and low rates of seedling survival, have caused populations to drop. Experiments by scientists in India and South Africa achieved germination rates of above 83 percent when seeds were pre-treated with hot water and then exposed to alternating light and dark conditions at 59°F (15°C). As these conditions are relatively straightforward to reproduce, local farmers should now be able to grow plants from seeds rather than relying on dwindling wild stocks.

LEONTOPODIUM ALPINUM

The white-flowering edelweiss, made famous in the film *The Sound of Music*, grows across Europe's mountains, from the Pyrenees to the Alps and onwards into the Balkans. It favors alpine meadows and steppe areas, at altitudes as high as 11,000 feet (3,350 m). Edelweiss populations have declined due to past collecting in the wild. The species is now considered to be critically endangered in the Ukrainian Carpathians and endangered in Bulgaria and Germany. It is protected in many countries.

1 *Able to absorb high levels of ultraviolet light, the edelweiss (*Leontopodium alpinum*), has potential as a sunscreen. However, it is one of several alpine plant species being pushed to higher altitudes by climate change.*

2 *Seeds of the rare Chilean blue crocus (*Tecophilaea cyanocrocus*) have been banked to conserve it for posterity.*

As well as being coveted by tourists and botanists for its star-shaped flowers, the edelweiss is used in cosmetics and to treat rheumatic pain. In Switzerland, which has adopted the edelweiss as its national flower, it is being selectively bred to produce a commercial crop. It also has potential as a suncream; scientists recently found that white hairs on its leaves completely absorb the high levels of ultraviolet light the plant is exposed to in its mountain habitats.

Climate change is already threatening wild populations of *L. alpinum*, however. The first pan-European study of changing mountain vegetation, published in 2012, found that cold-loving species, including edelweiss, are being pushed to higher altitudes at a faster rate than anticipated.

TECOPHILAEA CYANOCROCUS

This plant has striking cobalt-blue flowers, which gave rise to its common name, the Chilean blue crocus, but it is not a true crocus. It was first described in 1862 by the German botanist Friedrich Leybold and was believed to grow only in hills around Santiago, Chile, at between 6,500 and 9,800 feet (2,000 and 3,000 m) in altitude.

Overcollecting led to its disappearance and, from the 1950s, it was believed to be extinct. Then, in 2001, a population was discovered to be thriving near Santiago. Seeds from *T. cyanocrocus* are among those from 850 species of Chilean plants which have been collected and banked through a collaboration between the MSBP and Chile's Instituto de Investigaciones Agropecuarias. Specimens have been successfully cultivated at Kew Gardens, ensuring this rare plant's continued existence.

2

ARABICA COFFEE

(COFFEA ARABICA)

GENUS Coffea

FAMILY Rubiaceae

SEED SIZE Two seeds are contained in a cherry ⅜ to ⅝ inch (10–15 mm) across

TYPE OF DISPERSAL Animal (birds, small-and medium-sized mammals)

SEED STORAGE TYPE Intermediate

COMPOSITION Oil: 9.75%; Protein: 9.5%

Coffee is the world's most popular drink, with more than two billion cups drunk every day worldwide. So important is our daily pick-me-up that it is one of the world's major trading commodities; the industry supports at least 25 million coffee-farming families. Yet all may not be well for the future of the coffee industry. This is because the plants that supply our coffee have very limited genetic diversity, which means they are susceptible to climate change, pests, and diseases. Early plantations were often established from a small number of plants and in some cases a single plant, and this genetically poor stock went on to form the basis for global coffee production.

Each coffee seed has a pale outer shell, or parchment, which is enclosed by red, sweet juicy flesh. The red flesh and parchment are removed to provide the green beans that our coffee comes from. These are then roasted and ground. Although there are 124 coffee species, only two are important commercially: *Coffea canephora* (robusta) and *Coffea arabica* (Arabica). The best coffee comes from the latter. Arabica resulted from the hybridization of the two species *Coffea eugenioides* and *Coffea canephora* around 50,000 to one million years ago. An individual plant of *C. arabica* is able to produce fertile seed from its own pollen, which is almost unique among the 124 known species.

Today, the world's plantations of *C. arabica* probably contain less than five percent of the genetic variation that exists today in Ethiopia, where wild *C. arabica* plants grow. Those wild plants have the potential to be used to increase the genetic diversity of modern coffee crops, and indeed that has been happening on quite a regular basis over the history of Arabica cultivation. However, wild Arabica itself is also under threat from climate change. A study conducted by scientists at Kew Gardens and the Ethiopian Environment and Coffee Forest Forum concluded that climate change could wipe out wild Arabica coffee plants within 70 years. The study found that by 2080, climate change could reduce the extent of habitats favored by *C. arabica* in Ethiopia and South Sudan by between 65 and 100 percent.

Seeds from plants of the species *C. arabica* are not easy to bank, as they are "intermediate," partway between orthodox seeds that can be dried and stored and recalcitrant ones that cannot. They cannot, therefore, be stored in conventional seed banks under conditions of low temperature and moisture. In vitro germplasm collections, living collections, and cryopreservation are possible alternatives; however, these are costly and time-consuming methods to maintain. Research to find ways to preserve genetic material from *C. arabica* is vital if we are to conserve wild populations, and it may also help to underpin the coffee industry. If it fails, caffeine addicts will have to hope that taxonomists, botanists, and seed scientists are able to find a replacement among the 123 so far uninvestigated *Coffea* species. And if they cannot, humanity may learn a hard lesson about the importance of seeds and the value of conserving genetically diverse landraces, cultivars, and species.

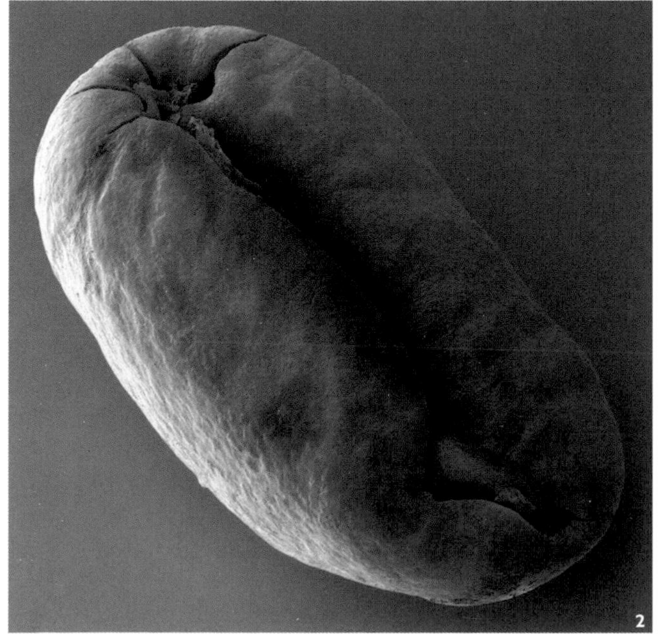

1 *Green coffee cherries of* Coffea arabica *ripen to red. The beans we roast are concealed inside the fruits.*

2 *A roasted coffee bean, from which comes the world's favorite drink.*

GLOSSARY

Anemochory Dispersal of seeds by wind, as in dandelion clocks, sycamore seeds, and tumbleweed.

Angiosperms All flowering seed plants, the most diverse group on earth, amounting to 352,000 species or 96 percent of all terrestrial vegetation.

Apomixis A process whereby plants reproduce asexually via seeds from the maternal tissue of the ovule, without the processes of meiosis or fertilization.

Chromosomes Structures made of DNA that carry genetic information (genes).

Clade A life-form group comprising an ancestor and all its lineal descendants. For example, all monocots form a single clade.

Cultivar A plant variety that has been deliberately cultivated and maintained for certain characteristics.

Dicots, or dicotyledons, refers to one of the two major groups of angiosperms. The seedlings of dicots have two seed leaves, or cotyledons, in the embryo of the seed. Dicots include most common garden plants, shrubs, and trees, as well as broad-leafed flowering plants, such as magnolias, roses, and geraniums.

Diploid Cells with two sets of chromosomes, one from each parent. Diploid cells, which include most organisms, reproduce through mitosis.

Eukaryote Any organism in which the cell or cells contain a nucleus and other organelles (cell compartments) enclosed within internal membranes.

Gametophyte The haploid multicellular stage in the life cycle of a plant arising through the alternation of generations.

Germplasm The seeds and other living genetic resources maintained for breeding, preservation and other research purposes. Seed banks store germplasm.

Gymnosperms A group of non-flowering plants that includes conifers, cycads, and gnetophytes. Together with the angiosperms they make up the spermatophytes or seed plants.

Haploid Cells with one set of chromosomes, called haploid, are created when a diploid cell divides through meiosis.

Hydrochory Dispersal of seeds by water. This can take the form of nautochory (seaborne), bythisochory (in flowing water), and ombrochory (raindrops).

Landrace A long-established domesticated plant that has adapted to thrive in specific local conditions, and thus may be able to endure the challenges posed by climate, pests, and disease.

Meiosis A process of cell division during which the number of chromosomes in each cell is halved from two (diploid) to one (haploid).

Mitochondria Mobile cytoplasmic organelles found in eukaryotes, they generate most of the cell's supply of adenosine triphosphate, a source of chemical energy.

Mitosis A process of cell division, occurring only in eukaryotic cells, whereby each of the daughter cells has a set of chromosomes identical to that of the parent.

Monocots, or monocotyledons, are one of the two major groups of angiosperms. The seedlings of monocots have only one seed leaf in the embryo of the seed. Monocots include orchids, daffodils, lilies, irises, palms, grasses, and sedges.

Paraphyletic A group with a common ancestor but, unlike a clade, only some of the subsequent lineal descendants.

Polyploidy Cells or organisms are referred to as polyploid when they have multiple sets of chromosomes, unlike diploid cells, which have two. Polyploidy arises through a failure in meiosis. It is a common phenomenon in plants, but rare in animals.

Prokaryotes Single-celled organisms without a membrane-bound nucleus, mitochondria, or organelles (cell compartments). These organisms, either bacteria or archaea, were the first forms of life on Earth.

Pteridophytes Major group of spore-producing vascular plants which do not bear seeds, such as ferns, clubmosses, and horsetails.

Sporophyte The diploid multicellular stage in the life cycle of a plant.

Zoochory Dispersal of seeds via animals through four main methods: being hoarded, being defecated or regurgitated, being taken underground (by ants, aka myrmechory), and in being attached to fur or feathers.

Zygote The eukaryotic cell formed by the fusion of two gametes is known as a zygote. It is the earliest developmental stage in multicellular organisms.

FURTHER READING

Abbott, R. J., Holderegger, R., Lewis Smith, R. I., and Stehlik, I. (May 2003) "Populations of Antarctic Hairgrass (*Deschampsia antarctica*) show low genetic diversity." *Arctic, Antarctic, and Alpine Research* Vol. 35, No. 2, 214-217. INSTAAR, University of Colorado.

Anderson, J., Saldaña Rojas, J., and Flecker, A. (2009) "High-quality seed dispersal by fruit-eating fishes in Amazonian floodplain habitats." *Oecologia* Vol. 161, Issue 2, 279-290.

The Angiosperm Phylogeny Group (2009) "An update of the Angiosperm Phylogeny Group classification for the orders and families of flowering plants: APG III." *Botanical Journal of the Linnean Society*, 161: 105–121.

Anjum, F., Begum, W., Rehman, H., and Tabasum, H. (2014) "*Rheum emodi* (Rhubarb): A Fascinating Herb." *Journal of Pharmacognosy and Phytochemistry* 3 (2) 89–94.

Arathi, H. S. (2012) "A Comparison of Dispersal Traits in Dandelions Growing in Urban Landscape and Open Meadows." *Journal of Plant Studies* Vol. 1, No. 2.

Bharali, S., Khan, M. L., Paul, A., and Tripathi, O. P. (2013) "Anthropogenic Disturbances Led to Risk of Extinction of *Taxus wallichiana* Zuccarini, an Endangered Medicinal Tree in Arunachal Himalaya." *Natural Areas Journal* 33(4): 447-454.

Bicknell, R. A. and Koltunow, A. M. (2004) "Understanding Apomixis: Recent Advances and Remaining Conundrums." *The Plant Cell* Vol. 16, suppl 1 S228-S245.

Buchman, S. L. (1987) "The ecology of oil flowers and their bees." *Annual Review of Ecology, Evolution, and Systematics* 18:343–69.

Bullock, J. M. and Moy, I. L. (2004) "Plants as seed traps: inter-specific interference with dispersal." *Acta Oecologica* 25: 35–41.

Chaudhary, P., Shrestha P., and Vernooy, R. (2013) "Community Seed Banks in Nepal: Past, Present, Future." *Proceedings of a National Workshop*, LI-BIRD/USC Canada Asia/Oxfam/The Development Fund/IFAD/Bioversity International, 14-15 June 2012, Pokhara, Nepal.

Chitale, V. S., Behera, M. D., Roy, P. S. (2014) "Future of Endemic Flora of Biodiversity Hotspots in India." *PLOS ONE* 9 (12).

Chittka, L. and Whitney, H. (August 2007) "Warm flowers, happy pollinators." *Biologist*, Vol. 54, No. 3. 154-159.

Cowell, C. "Protocols for Fynbos Restoration." 3rd Global Botanic Gardens Congress. (http://www.cabi.org/isc/FullTextPDF/2009/20093081924.pdf)

Crane, P. R. and Leslie, A. B. (2013) "Major events in the evolution of plant life." In: Losos, J. et al. (eds), *The Princeton Guide to Evolution*, 143-151, Princeton University Press.

Davidson, A. (2006) *The Oxford Companion to Food*. (2nd edition) Oxford University Press.

Dempewolfe, H. et al. (2014) "Adapting Agriculture to Climate Change: A Global Initiative to Collect, Conserve, and Use Crop Wild Relatives." *Agroecology and Sustainable Food Systems*, 38:369–377.

Deza, E. and Deza, M. M. (2014) *Encyclopedia of Distances* Springer.

Prehistoric: The Definitive Visual History of Life on Earth (2009) Dorling Kindersley Ltd.

Dumais, J., Evangelista, D., and Hotton, S. (2011) "The mechanics of explosive dispersal and self-burial in the seeds of the filaree, *Erodium cicutarium* (Geraniaceae)." *Journal of Experimental Biology*, 214 (4): 521 DOI: 10.1242/jeb.050567

Ekué, M. R. M. et al. (2010) "Uses, traditional management, perception of variation and preferences in ackee (*Blighia sapida* K.D. Koenig) fruit traits in Benin: implications for domestication and conservation." *Journal of Ethnobiology and Ethnomedicine* 6: 12.

Fenner, M. and Thompson, K. (2005) *The Ecology of Seeds* Cambridge University Press.

Food and Agriculture Organization of the United Nations (2009) International Treaty on Plant Genetic Resources for Food and Agriculture. FAO.

Forest, F. et al. (2014) "Pollinator shifts as triggers of speciation in painted petal irises (*Lapeirousia: Iridaceae*)." *Annals of Botany* 113: 357–371.

Fry, C., Seddon, S., and Vines, G. (2011) *The Last Great Plant Hunt; The Story of Kew's Millennium Seed Bank* Kew Publishing.

Fry, C. and Willis, K. (2015) *Plants from Roots to Riches* John Murray.

FURTHER READING

Gebser, R. and Matt, F. (2011) "Biodiversity decline can increase the spread of infectious diseases." TEEBcase, available at: www.TEEBweb.org.

Godoy, J. A. and Jordano, P. (2001) "Seed dispersal by animals: exact identification of source trees with endocarp DNA microsatellites." *Molecular Ecology* 10 (9): 2275-83.

Gottfried, M. et al. (2012) "Continent-wide response of mountain vegetation to climate change." *Nature Climate Change* 2, 111–115.

Goulson, D., Hawson, S. A., and Stout, J. C. (1998) "Foraging bumblebees avoid flowers already visited by conspecifics or by other bumblebee species." *Animal Behaviour* 55(1): 199-206.

Graz, F. P. (2002) "Description and ecology of *Schinziophyton rautanenii* (Schinz) Radcl.-Sm in Namibia." *Dinteria* 27: 19-35. Windhoek, Namibia.

Hodges, M. (2010) "Nature's Back-up: The Global Seed Vault." *Wired* (http://www.wired.co.uk/magazine/archive/2010/11/features/natures-backup).

Honnay, O. et al. (2010) "Patterns of population genetic diversity in riparian and aquatic plant species along rivers." *Journal of Biogeography* 37 (9): 1730-9.

International Food Policy Research Institute. (2002) *Green Revolution: Curse or Blessing?*

Jan de Boer, H. et al. (2012) "A critical transition in leaf evolution facilitated the Cretaceous angiosperm revolution." *Nature Communications* 3, Article number: 1221.

Jansen, P. A., Kays, R., Knecht, E. M. H., Vohwinkel, R., and Wikelski, M. (2011) "The effect of feeding time on dispersal of Virola seeds by toucans determined from GPS tracking and accelerometers." *Acta Oecologica*.

Johansson, M. E., Nilsson, C. and Nilsson, E. "Do rivers function as corridors for plant dispersal?" *Journal of Vegetation Science* 7: 593–598, 1996.

Kelly, D., Nelson, X. J., and Young, L. (2012) "Alpine flora may depend on declining frugivorous parrot for seed dispersal." *Biological Conservation* 147, 133–142.

Lack, A., Proctor, M., and Yeo, P. (1996) *A Natural History of Pollination* Collins; 2nd Revised edition.

Lane, N. (December 2002) "The Big O." *New Scientist*. (https://www.newscientist.com/article/mg17623735.500-the-big-o/#.VAdSC2RdVTp)

Larson, D. F., and Otsuka, K. (Eds.) (2013) *An African Green Revolution: Finding Ways to Boost Productivity on Small Farms*. Springer Science+Business Media.

León-Lobos, P., Pritchard, H.W., Sandoval, A., and Way, M. (2010) "The contribution of the Millennium Seed Bank Project to ex situ plant conservation in Chile." *Kew Bulletin* Vol. 65, Issue 4, 595-601.

Luskotov, I. G. (1999) *Vavilov and his Institute. A history of the world collection of plant genetic resources in Russia* International Plant Genetic Resources Institute (http://www.vir.nw.ru/files/pdf/books/Vavilov_and_his_institute.pdf)

Mackenzie, A. (July 2014) "Australian PlantBank." *Architecture Australia*.

Madison, D. (Foreword) (2008) *Edible: An Illustrated Guide to the World's Food Plants* National Geographic.

Malthus, T. (1798) "An essay on the principle of population, as it affects the future improvement of society with remarks on the speculations of Mr. Godwin, M. Condorcet, and other writers." London, printed for J. Johnson, in St. Paul's church-yard. (http://www.gutenberg.org/files/4239/4239-h/4239-h.htm)

Marshall, M. (October 2013) "Hunter-gatherers got on fine with Europe's first farmers." *New Scientist* (https://www.newscientist.com/article/dn24385-hunter-gatherers-got-on-fine-with-europes-first-farmers)

McElwain, J. and Willis, K. (2013) *The Evolution of Plants* Oxford University Press. 2nd edition. 2013.

Moeller Gorman, R. "Cooking up bigger brains." *Scientific American* (http://www.scientificamerican.com/article/cooking-up-bigger-brains/?page=2)

Muir, J. (1992) *The Eight Wilderness Discovery Books* Diadem Books.

Nabhan, G. P. (2009) *Where Our Food Comes From: Retracing Nikolay Vavilov's Quest to End Famine* Island Press.

Nagy, L. and Grabherr, G. (2009) *The Biology of Alpine Habitats* Oxford University Press.

Nelson, E. C. (2000) *Sea beans and nickar nuts* BSBI Handbook no. 10. Botanical Society of Britain and Ireland.

Neng, C. Y. (2010) "Orchid smugglers and the use of biotechnology to combat them." *AsPac Journal of Molecular Biology and Biotechnology* Vol. 18 (1): 175-179.

Newton, R. J., Bond, W. J., and Farrant, J. M. (2006) "Effects of seed storage and fire on germination in the nut-fruited Restionaceae species, *Cannomois virgata*." *South African Journal of Botany* 72, 177-180.

Niveditha, V. R., and Sridhar,K. R. (June 2014) "Nutritional and bioactive potential of coastal sand dune wild legume *Canavalia maritime* (Aubl.) Thou. – An overview." *Indian Journal of Natural Products and Resources* Vol. 6. (2), 107–120.

Padulosi, S., Hammer, K., and Heller, J. (Eds.) "Hulled Wheat." *Proceedings of the First International Workshop on Hulled Wheats* 21-22 July 1995, Castelvecchio Pascoli, Tuscany, Italy.

Parker, L. and Torr, G. (2010) "The Seed Seekers." *Australian Geographic* Issue 90.

Redshaw, C.H. and Schmidt, W. (February 2015) "Evaluation of biological endpoints in crop plants after exposure to non-steroidal anti-inflammatory drugs (NSAIDs): Implications for phytotoxicological assessment of novel contaminants." *Ecotoxicology and Environmental Safety* Vol 112, 212–222.

Shirihai, H. (2002) *A Complete Guide to Antarctic Wildlife: The Birds and Marine Mammals of the Antarctic Continent and the Southern Ocean* Alula Press Oy, Finland.

Telewski, F. W. and Zeevaart, J. A. D. (2002) "The 120-year Period for Dr Beal's Seeds Viability Experiment." *American Journal of Botany* 89(8) 1285–1288.

Thomas, J. A. et al. (March-April 2010) "Corruption of ant acoustical signals by mimetic social parasites: Maculinea butterflies achieve elevated status in host societies by mimicking the acoustics of queen ants." *Communicative and Integrative Biology* 3(2): 169–171.

Traveset, A. and Willson, M. F. (2000) "Chapter 4: The Ecology of Seed Dispersal" in *Seeds: the Ecology of Regeneration in Plant Communities*. (Ed. M. Fenner). 2nd Edition. CAB International.

Warren, R. et al. (July 2013) "Quantifying the benefit of early climate change mitigation in avoiding biodiversity loss." *Nature Climate Change* Vol 3.

PICTURE CREDITS

The publisher would like to thank the following individuals and organizations for supplying images.

Alamy/Les Gibbon: 92C; Andrea Jones Images: 128; National Geographic Image Collection: 98.

Biodiversity Heritage Library: 20, 36, 51, 55, 65L, 65R, 84, 87.

Corbis/© Denis Sinyakov/Reuters: 148.

Flickr.com/Alvesgaspar: 86T; Ccotton: 120B; Michael Clarke Stuff: 120R; Dorena-WM: 109; Dag Terje Filip Endresen: 33T, 34T, 35B; Goldentakin: 121T; Sodai Gomi: 93C; Luigi Guarino: 31, 32, 33B, 34B, 35T; Vishruth Harithsa: 179; Matthias Heyde/Landbruks- og matdepartementet: 164; Matt Lavin: 147; Magnus Manske: 183; Neil Palmer: 174, 175, 176, 177; Kristine Paulus: 11, 124, 125, 126, 127; Plantsurfer: 46R; Liam Quinn: 99; David Thou: 159.

FLPA/Mark Moffett/Minden Pictures: 89, 93L.

The Germplasm Bank of Wild Species, Kunming Institute of Botany, Chinese Academy of Sciences: 94, 95, 96, 97.

Getty Images/Buena Vista Images: 131; De Agostini Picture Library: 46L; Darlyne A. Murawski/National Geographic: 13.

© Martin Hamilton: 130B, 130T. © Fred Henze: 92R.

Hippocampus Bildarchiv/Frank Teigler: 41.

© Tony Kirkham: 73. © Paul Latham: 66.

Library of Congress, Washington, D.C.: 30.

© Augustin Konda ku Mbuta: 67.

Michigan State University/Kurt Stepnitz: 145.

Michigan State University Archives and Historical Collections: 144.

Nature Picture Library/Sue Daly: 120T; Fabrice Cahez: 85; Pete Oxford: 112, Jose B. Ruiz: 136; Visuals Unlimited: 111; Wegner/ARCO: 138; Wild Wonders of Europe/Möllers: 157; Bert Willaert: 64; Jeff Wilson: 100; Konrad Wothe: 118B. NASA: 167L.

© Rosemary Newton: 142, 143. © Helen Pickering: 133.

Royal Botanic Gardens, Kew: 50, 139, 140, 153, 154, 155, 168; Photo by John Dickie: 152; Photo by Andrew McRobb: 38, 73 (inset), 92L, 116, 141; Photo © The National Archives: 149L.

Royal Botanic Garden Sydney: 69, 70, 71; Photo by John Gollings: 68.

Science Photo Library/George Bernard: 48; Thierry Berrod, Mona Lisa Production: 80T; Dr Jeremy Burgess: 77L, 80BR; Nigel Cattlin: 23T; Garry Delong: 76; Stefan Diller: 13T (inset); Eye of Science: 23B; Bob Gibbons: 78B; Pascal Goetgheluck: 16; Geoff Kidd: 181; Frans Lanting/Mint Images: 53; Leonard Lessin: 88; London School of Hygiene & Tropical Medicine: 167R; Raul Martin/MSF: 17; Natural History Museum, London: 49; Claude Nuridsany & Marie Perennou: 77R, 119; Power and Syred: 54, 57B, 80BL, 137, 146 (inset), 185B; Noble Proctor: 78T; Philippe Psaila: 118T; Dr. Peter Siver/Visuals Unlimited: 44; Sinclair Stammers: 47; Dr Keith Wheeler: 21, 59.

Shutterstock/Alslutsky: 171; ChameleonsEye: 27L; Colette3: 146; Ethan Daniels: 115; Eugalo: 93R; Eye-blink: 57T; Martin Fowler: 170L; Irina Fuks: 170R; HartmutMorgenthal: 103B; Aleksandr Hunta: 182; JuRitt: 25; Karen Kaspar: 24; Keneva Photography: 86B; Saied Shahin Kiya: 107 (inset); D. Kucharski K. Kucharska: 103T; Holly Kuchera: 120L; Steve McWilliam: 79; Mimohe: 165; Papa Bravo: 90; Alexander Piragis: 101BL; Ruzanna: 26L; Aleksey Sagitov: 101BR, 101T; Solarseven: 107; Stasis Photo: 185T; Stanislaw Tokarski: 26R; Sergey Uryadnikov: 121B; James Whitlock: 27R; Yankane: 27C.

© Wolfgang Stuppy: 39, 133BL, 133BR. © Marilyn Tyzack: 169.

© Universiteit van Amsterdam: 45.

U.S. Department of Agriculture: 103T (inset).

U.S. Fish & Wildlife Service/G. Wallace: 129.

Wikimedia Commons/Roger Culos: 66 (inset); Ton Rulkens: 156; Marco Schmidt: 37; Scott Zona: 52.

Every effort has been made to acknowledge the pictures, however the publisher apologizes for any unintentional omissions.

INDEX

A

acacia 36–37, 149
ackee 65
Adapting Agriculture to Climate Change 40, 178
algae 44–45, 60–61, 63, 79
alpine seeds 180–82
amaranth 129, 178
Angiosperm Phylogeny Group (APG) 151
angiosperms 11–12, 53, 56, 58–63, 77, 79, 82–84, 86, 150–51
annuals 58–59, 138
asexual reproduction 81
Asomaning, Joseph 66
Azuma, Akira 117

B

Baldock, Sam 170
ballistic propulsion 119
Bandurski, Robert 144
Banks, Joseph 156
Barcoding Chinese Plants Project 96
Baskin, Carol 137, 158
Baskin, Jerry 137, 158
Beal, William J. 144–45
bees 88, 91, 93, 166
Botanical Gardens 72, 111, 137, 144–45, 165
boxwood 128
Brighton and Hove Council 169–70
brush harvesting 169–70
Bureau of Land Management 124

C

Campos-Arceiz, Ahimsa 112–13
carats 29
Carter, Howard 139
Caum, Edward L. 129
Cavendish plantations 24
centers of origins 31
Centre for Conservation and Research 113
Centre for Ecology and Hydrology 108, 117
Centre for Legumes in Mediterranean Agriculture 40
CGIAR 163, 174, 179
Chanyenga, Tembo F. 66
Chapman, Ted 169
Chase, Mark 58–59
China Plant Specialists Group 96
Chinese Academy of Sciences 96
civilization 10, 16, 59
climate change 22, 24–25, 40, 50, 53–54, 56, 108, 114, 157, 165, 177–81, 184
coastal seeds 156–57
coconuts 115, 131
coffee 93, 166, 184–85
Commonwealth Environment Protection and Biodiversity Conservation Act 72
Conservation and Restoration Group 158
Cookson, Isabel 46
Corner, E. J. H. 79
Crane, Peter 45, 47, 50, 54
crop wild relatives 25–27, 31, 162–63, 178, 184
crop yields 19, 21–22, 25
cryopreservation 39
cucumbers 117, 119
cycads 51

D

Darlington, Henry 144
Darwin, Charles 11, 54, 58, 87, 115
Daws, Matt 149
Debouck, Daniel 174–77
Department of Applied Botany and Plant Breeding 30–35
devil's claws 38–39
Dezhu Li 97
dispersal 104–33
DNA 47, 53, 58, 77, 89, 95–97, 108, 147, 151
dormancy 136–39, 141, 143–45, 155, 158
double fertilization 76–83
drift seeds 115

E

Eastwood, Ruth 24–25, 31
Economic Botany Collection 139
edelweiss 181–82
Eden Project 92
enrichment planting 39
Environment and Coffee Forest Forum 184
Esmée Fairburn Foundation 169
Etrich, Igo 117
evolution 10–13, 16–17, 19, 25, 42–73, 76–77, 79, 82, 85–86, 89, 106, 111–13, 116, 137
extinction 56, 62, 64, 67, 72, 108, 112, 128, 148, 153, 157, 180

F

Fernando, Prothiviraj 113
Fertile Crescent 19
Floristic Center 65
flowering plants 52–57, 62, 76–81, 98–100, 108, 150
Food and Agriculture Organization (FAO) 39, 162–63
Forest, Félix 89
fossils 46–47, 51, 54, 72
Fournel, Jacques 92
future foods 178–79

G

gap analysis 24
Gapper, John 170–71
genetic diversity 22, 24, 26, 30, 54, 56, 81, 88–89, 91, 99, 108, 114, 142, 147, 162–63, 165–70, 184
Genetic Resources Program 174
germination 134–59, 181
Germplasm Bank of Wild Species (GBOW) 94–97, 102
Global Crop Diversity Trust 24, 163, 175, 178
grass pea 40
gravity dispersal 119
Green Revolution 22
Greenbelt Native Plant Center (GNPC) 11, 122–27
groundnuts 178
gymnosperms 11, 53–54, 58, 62–63, 77, 79, 81–83, 86

H

hair grass 99
Hiscock, Simon 79, 84, 88–89, 91
hominids 16–17
Hooker, William 65
Howard Hughes Medical Institute 81

I

iFlora 97
inkberry 156–57
Instituto de Investigaciones Agropecuarias 182
International Center for Agricultural Research in Dry Areas (ICARDA) 40, 163
International Center for Tropical Agriculture (CIAT) 24, 174–77
International Crop Research Institute 179
International Maize and Wheat Improvement Center (CIMMYT) 175
International Union for Conservation of Nature (IUCN) 66, 96, 102, 128–29, 180
island seeds 128–31

J

jackfruit 178
Jodrell Laboratory 58
John Innes Horticultural Institute 30

K

Kenya Agricultural Research Institute 179
Kew Gardens 24–25, 31, 36–37, 51, 58, 64–65, 72, 89, 92–93, 111, 128–29, 132,

137, 139–40, 149, 152–55, 157, 168, 177, 182, 184
Kivilaan, Aleksander 144
Kunming Institute of Botany 96

L

Larkin, David 169–70
Lawton, John 168–69
Lenin, Vladimir 31
Leybold, Friedrich 182
life cycles 77, 79, 88–89, 150

M

Malthus, Thomas 19
Masson, Francis 51
melons 38
Micheneau, Claire 92
Mid-Atlantic Regional Seed Bank (MARS-B) 124–25
Millennium Commission 153
Millennium Ecosystem Assessment 12, 165
Millennium Seed Bank Partnership (MSBP) 24, 25, 31, 36–37, 39, 66–67, 69, 72, 99, 102, 111, 120, 130, 132, 137, 140–41, 149, 152–55, 157–58, 165, 168–70, 177–78, 181–82
millet 179
mongongo 132–33
Monsanto 22
Montserrat National Trust 130
Moscow Agricultural Institute 30
Muir, John 100
Mulanje cedar 66–67
mustard 131

N

National Center for Agricultural Research and Extension (NCARE) 37
National Institute for Basic Biology 45
National Museum of Wales 99
National Tree Seed Center 132, 155
Native Seed Hub Initiative 168–69
natural selection 10–11, 51, 76, 84, 86
Nature Improvement Areas (NIA) 169–71
nectars 87–88, 91–93
Nesbitt, Mark 139
Newton, Rosemary 137, 140, 142–43
Nikolaeva, Marianna 137
Noble, David 72

O

oil palms 67
Okuno, Yoshinori 117
orchids 59, 70, 79, 87, 91–92, 109, 150
orphan crops 179
orthodox seeds 39, 65, 136, 184

P

parachuting 106, 117
pearlwort 99
perennials 59, 81, 131
photosynthesis 45, 48, 51, 56
phylogeny 62–63, 89, 151
pines 72–73
Plant Genetic Resources for Food and Agriculture, Treaty on 22, 24
plant-animal relationships 51, 54, 85–89, 91–93, 100, 106, 110–15, 120–21, 147, 171
PlantBank 68–71
pollination 51, 54, 56, 79, 81, 84–89, 91–93, 106, 166
polyploidy 58, 77
poppies 100, 119, 145
pre-breeders 24
pribby 130
Probert, Robin 158
Purcell, William F. 167

R

rainforests 64–71
recalcitrant seeds 39, 65–66, 136, 158, 184
Red List of Threatened Species 66
reproduction 74–103
resilience 24, 26–27, 59, 81, 177–79
rewilding 113
rhizomes 81
rice 27, 163, 178
Rich, Tim 99
Rio Earth Summit 162
Roundup 22

S

Sacandé, Moctar 67, 132
salt bush 157
saxifrage 180
seed banks 12–13, 24, 31–36, 39, 68–71, 94–97, 122–27, 140–41, 147, 152–55, 158, 160–87
Seed Information Database (SID) 141
Seeds of Success (SOS) 124–25
Smith, Paul 111, 165
Smithsonian Institute 108, 113
Solanaceae Germplasm Bank 165
South Downs National Park 169–70
speciation 89
Species Survival Commission 96
spore-bearing plants 44–48, 50, 76–77
Stalin, Joseph 31
Stanmer Nurseries 169–70
stinking hawksbeard 157
stonecrops 58
sunflowers 26
Svalbard Global Seed Vault 163, 164, 175
sycamores 116–17

T

Teerlink, Jan 149
teff 179
Telewski, Frank 144–45
Thompson, Ken 106, 108
Threatened Species Conservation Act 72
tobacco 59
tomatoes 26
tooth decay 17
Toth, Edward 123–27
tumbleweeds 118
Tutundjian de Cartavan, Christian 139

U

Unilever 67
universities 24, 30, 45, 47, 65–66, 79, 91, 93, 106, 108, 112, 137, 139, 144, 156, 165, 181
Useful Plants Project 37, 155

V

Van Gelder, Roelof 149
Vavilov Institute of Plant Genetic Resources (VIR) 31–35
Vavilov, Nikolai 30–35
Vielle-Calzada, Jean-Philippe 81

W

Wakehurst Place 72, 154
water dispersal 114–15
water lilies 92
water pollination 84
Wellcome Trust 153–54
Wels, Franz X. 117
wheat 27, 163, 178
wind dispersal 116–18
wind pollination 84–85
windflowers 100
wood anemone 158–59
WorldClim 140
Wu Zhengyi 94–95
WWF 112

Y

yew 102–3

ACKNOWLEDGMENTS

The author would like to thank the following people for their help during the production of this book:

Aaron Davis; Cicely Henderson; Clare Trivedi; Craig Brough and the library team; David Goyder; Félix Forrest; Georgina Hills; Gina Fullerlove; Herta Kolberg; John Dickie; Kate Hardwick; Mark Chase; Mark Nesbitt; Martin Hamilton; Michiel van Slagaren; Moctar Sacandé; Richard Wilford; Robin Probert; Rosemary Newton; Ruth Eastwood; Ted Chapman; Tim Pearce; Tony Kirkham; William Baker; and Wolfgang Stuppy (all The Royal Botanic Gardens, Kew).

Ahimsa Campos-Arceiz (University of Nottingham, UK); Cathy Offord (The Royal Botanic Gardens and Domain Trust, Australia); Daniel Debouck, Nathan Russell and Neil Palmer (International Center for Tropical Agriculture, Colombia); David Larkin (Brighton and Hove City Council, UK); Edward Toth (Greenbelt Native Plant Center, New York City Department of Parks and Recreation, USA); Frank Telewski (Michigan State University, USA); Hannes Dempewolf (Global Crop Diversity Trust, Germany); Igor Loskutov (N. I. Vavilov Institute of Plant Genetic Resources, Russia); Ken Thompson (University of Sheffield, UK); Khaled Abulaila (National Center for Agricultural Research and Extension, Jordan); Lydia Guja (Australian National Botanic Gardens and Centre for Australian National Biodiversity Research, CSIRO); Paul Smith (Botanic Gardens Conservation International, UK [formerly Kew's MSBP]); Sir Peter Crane (Yale University, USA); Simon Hiscock (University of Bristol, UK); and Xiangyun Yang (Germplasm Bank of Wild Species, Kunming Institute of Botany, China).

Artwork references

The author and publishers are grateful for permission to use the following source material as references for artwork:

p.62 Figure 6.2 from "The Influence of Paleozoic Ovule and Cupule Morphologies on Wind Pollination" by K. J. Niklas in *Evolution* Vol 37, No 5 (1983). © 1983 John Wiley & Sons, Inc. With permission of Blackwell Publishing Ltd.

p.63 Data from "Major events in the evolution of plant life" by P. R. Crane and A. B. Leslie in *The Princeton Guide to Evolution* ed J Losos (Princeton University Press, 2013).

p.109 Figure 1b from "Global patterns in seed size" by Angela T. Moles et al. *Global Ecology and Biogeography*, Vol 16, Issue 1, (January 2007). © 2006 John Wiley & Sons, Inc. With permission of Blackwell Publishing Ltd.

p.117 Figure 2 from "Long distance seed dispersal by wind: measuring and modelling the tail of the curve" by J. M. Bullock & R. T. Clarke, *Oecologia* Vol 124, Issue 4, p.511 (September 2000). © Springer-Verlag 2000. With permission of Springer Science+Business Media

p.162-3 Data from "Our Dwindling Food Variety" National Geographic (http://ngm.nationalgeographic.com/2011/07/food-ark/food-variety-graphic). Source: Rural Advancement Foundation International.